AF364479

Beyond Cellulose: The Complex Biology of Plant Cell Walls

kinky

Copyright © 2024 by kinky

All rights reserved. No part of this book may be reproduced in any manner whatsoever without written permission except in the case of brief quotations embodied in critical articles and reviews.
First Printing, 2024

Table of Contents

Chapter 1: Plant Cell Walls

Biological Importance and Economic Significance

Plant cell walls are unique structures with dynamic abilities to withstand large amounts of physical force, expand, contract and undergo reconstructions to meet cellular demands. They are made of biopolymers that have evolved qualities that can withstand and resist external attack by microbes and insects, while allowing for growth, making them tough and flexible (Anderson and Kieber, 2020). Despite being restricted to a fixed position at the site of germination, plants regulate their growth in ways that allow deposition of new materials into the existing wall during growth, without which wall thinning and subsequent mechanical failure can occur (Cosgrove, 2014). Therefore, regulatory mechanisms that balance wall extensibility with wall deposition are central to plant growth and developmental processes (Wolf, 2017).

Plant cell walls are structurally complex heteropolymers with dynamic biological networks composed of interacting sugars, proteins, and polyphenolics, that provide support and protection to plants and determine both their morphological and mechanical properties throughout plant life. As the most abundant repository of biomass on Earth, plant cell walls are used as sources of food, animal feed, clothing, shelter and energy (Anderson and Kieber, 2020). The potential of plant cell wall materials to be converted into biofuels to replace fossil fuels has increased their significance (Ragauskas et al., 2014). Similarly, plant cell walls constitute the key structural components of plants and many plant-based foods. They are well known for contributing to a range of "quality" characteristics, from organoleptic texture to the properties of dietary fiber (Waldron et al.,

2003). Pectins enhance the digestive quality of many foods by providing soluble and insoluble fibre (Lara-Espinoza et al. 2018) while in food science, pectin serves in texture control, thickening, emulsification, and stabilization (Leroux et al., 2003). Furthermore, several hydrolases, cellulases, endo-β-glucanases are produced by rumen microbiota to improve fiber digestibility of forage grasses in ruminants (Reddy et al., 2005) and the digestibility of these forages is dependent on the level of lignification. Over the years, Gum Arabic, which is a member of arabinogalactan-protein (AGP) family, has proven useful in food and biomedical industries. For example, its emulsifying, stabilizing, binding and shelf-life enhancing characteristics made it useful in the food industry, while Gum Arabic's anti-microbial, anti-inflammatory and anti-coagulant characteristics has fueled its interest and use in the pharmaceutical industry (Patel and Goyal, 2015).

Despite its biological and economic importance, much remains to be discovered about how plant cell walls components – cellulose, hemicellulose, pectins and cell wall proteins – are organized and assembled to form remarkable structures that stabilize plant cell wall architecture. Plants harbor more than 40 types of cells (Farrokhi et al., 2006) and each cell type possesses a distinct cell wall composition and organization. Remarkably, research efforts geared towards elucidating the biosynthetic pathways involved in cell wall biogenesis have identified many genes related to cellulose synthesis (Liu et al., 2015), matrix polysaccharide production (Persson et al., 2007), wall polymer modification and deposition (Kang et al., 2019), presenting an opportunity to better understand the underlying molecular mechanisms involved in the synthesis, assembly, integrity sensing and remodeling of plant cell walls. This brings us a step closer towards utilizing the tools of genetic engineering, cell and molecular biology and biochemistry to manipulate plant cell wall genes/proteins for improved biomass and bioenergy production.

Fundamentals of Plant Cell Wall Composition, Structure and Biosynthesis

Cellulose

Cellulose is the dominant mechanical component of plant cell walls with tensile strength properties, comparable to that of steel (Anderson and Kieber, 2020). Structurally, cellulose is comprised of multiple β-1,4- linked glucan chains synthesized at the cell surface by cellulose synthase (CESA) complexes (CSCs), and are hydrogen bonded to form cellulose microfibrils (**Figure 1-1**) (Kieber and Polko, 2019). Cellulose microfibrils are the main load-bearing component of the wall, and the orientation of the microfibrils determines the directionality of cell elongation and expansion (Rui and Dinneny, 2019). Considering the simplicity of its chemical nature, its assembly appears complex due to multiscale cellulosic microfibrils characterized by heterogenous mixture of highly crystalline regions interspersed with amorphous stretches or layers (Zhang et al., 2021). Cellulose microfibrils contain 18-24 chains (Thomas et al., 2015) and have a fiber diameter of 3 nm in primary walls, while cellulose in secondary walls coalesces into macrofibrils of about 20 nm in diameter, with the degree of polymerization for the cellulose glucan chains ranging from few hundreds of monomers to over 10,000 (Anderson and Kieber, 2020).

Figure 1-1

http://www1.biologie.uni-hamburg.de/b-online/library/webb/BOT311/PlantCellWalls00/PCW00-10.htm

Cellulose Production in the Plant Cell Wall

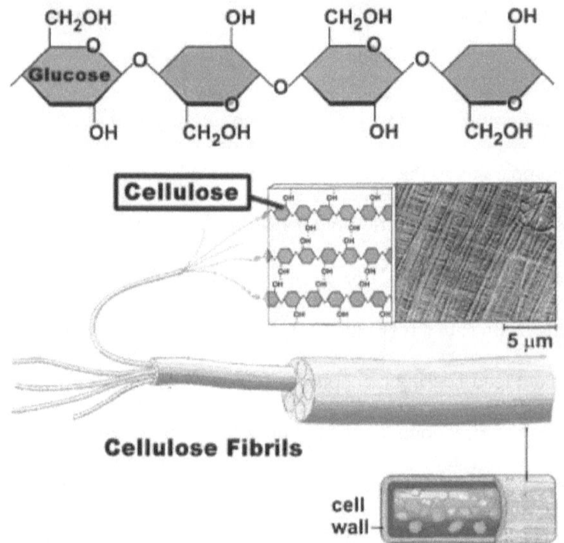

Note. Cellulose is made from simple monomers of β-1,4-linked glucose units, which aggregate to form cellulose microfibrils. Cellulose is the main constituent of plant cell walls.

Cellulose is synthesized by a group of enzymes known as cellulose synthases (CESAs) organized to form a protein complex called the cellulose synthases complex (CSC) in the plasma membrane (Mutwil et al., 2008). Each CSC subunit plays active roles in the glucan chain elongation using cytosolic uridine diphosphate-glucose (UDP-glucose) as a substrate (Anderson and Kieber, 2020). At least three different CESA isoforms are likely required to form a functional CSC (Taylor et al., 2003), which makes CESAs an integral membrane protein encoded by multigene families. Based on stoichiometric, microscopic and modeling evidence, CESA proteins form functional CSC in form of hexamers of trimers in the plasma membrane and represents an important step in understanding cellulose formation in plant cell walls (Hill et al., 2014).

Most of the functions of CESAs in plant growth have been unraveled mainly through genetic studies. Three Arabidopsis CESAs (CESA1, CESA3, and either CESA2, CESA5, CESA6, or CESA9) form a multimeric complex via protein-protein interactions to form cellulose in the primary cell wall (Arioli et al., 1998), while Arabidopsis CESA4, CESA7, and CESA8 assemble to produce cellulose during secondary cell wall biosynthesis (Taylor et al., 2003). Interestingly, these partially redundant CESAs involved in primary cell wall synthesis appear to have distinct functions as neither *CESA2* nor *CESA5* driven by a *CESA6* promoter fully complements a *cesa6* mutant (Desprez et al., 2007). Mutations in these *CESAs* have resulted in brittle culms, dwarfism and collapsed xylem tissues accompanied with significant growth abnormalities (Zhang et al., 2009). For example, a temperature sensitive allele mutant with an amino acid mutation in the catalytic domain of CESA1 (*rsw-1*) showed less crystalline cellulose, thin epidermal walls in the embryo, and a swollen root phenotype (Williamson et al., 2001). Surprisingly, a *CESA1* T-DNA knockout mutant was lethal and a similar observation was reported for a *CESA3* T-DNA knockout mutant. However, mutants heterozygous for *CESA1* and *CESA3* showed defective pollen phenotype (Persson et al., 2007b). Contrary to CESA1 and CESA3, loss of function of CESA2 displayed phenotypes that were comparable to the WT except for a shorter hypocotyl phenotype which was further exacerbated in *cesa2cesa6* mutants (Persson et al., 2007b). Other accessory proteins that regulate cellulose production have been identified and characterized, especially those that facilitate CSC trafficking and movements. For example, Arabidopsis COBRA is reported to be involved in cellulose assembly by binding to cellulose and impacting the crystallization of cellulosic microfibrils (Liu et al., 2013). Similarly, a genetic deficiency of Arabidopsis COBRA, rice Brittle Culm1 or maize Brittle Stalk2, severely reduces cellulose content (Sindhu et al., 2007). Cellulose synthase interactors (CSI1-3) were discovered to couple CSCs to cortical microtubles (CMT) (Gu et al., 2010), which can be uncoupled by cellulose synthase microtubule uncoupling (CMU) proteins (Liu et al., 2016b), the actions of which are modulated by the kinesin protein Fragile fiber1 (FRA1) (Ganguly et al., 2020). Other proteins such as AtCTL1 and AtCTL2 (Sanchez-Rodriguez et al., 2012), rice BC15/OsCTL1 (Wu et al., 2012), and maize BK4/ZmCTL1 (Jiao et al., 2019), as well as the CSC-associated protein KORRIGAN, a glycosyl hydrolyase 9 (GH9) protein (Vain et al., 2014), have been implicated in cellulose production but the underlying mechanisms of action remain elusive.

Hemicelluloses

Hemicelluloses which are comprised of xylans, xyloglucans, mannans, glucomannans, and β-(1,3;1,4)-glucans. All harbor β-(1,4)-glycosyl linked backbones, and most hemicellulosic backbones are synthesized by cellulose synthase-like proteins (CSLs) from the GT2 family (Scheller and Ulvskov, 2010) except for xylans, whose backbones are produced by Type II membrane proteins from the GT47 and GT43 families (Wu et al., 2010).

Xyloglucans. Xyloglucans comprise a heterogenous collection of polysaccharides with variable main and side chain lengths (Schultink et al., 2014). In many dicots, hemicellulose contains approximately 20% of the dry mass of primary cell walls (Schultink et al., 2014), but the xyloglucan content may be as low as 2% (Thimm et al., 2002). Xyloglucans consist of a β-(1,4) glucan backbone that is partly substituted with xylosyl residues linked to glucose at the *O*-6 position (**Figure 1-2**) and may be further modified by other side chain sugars whose substitution and sugar identity varies considerably (Tuomivaara et al., 2015). The cooperative biosynthetic mechanism was elucidated when radiolabeled UDP-[^{14}C]Glc and UDP-[^{14}C]Xyl were used as substrates and added to pea microsomal membranes and detectable radioactivity in β-1, 4-glucose backbones was observed. The addition of either UDP-[^{14}C]Glc or UDP-[^{14}C]Xyl did not result in detectable radioactivity in β-1, 4-glucose backbones, indicating that the incorporation of glucose is coupled with the synthesis Xyl side chain formation (Camirand and Maclachlan, 1986). Further investigation in sugar side chain incorporation revealed that the addition of either GDP- [^{14}C]Fuc or UDP-[^{14}C]Gal can incorporate Fuc or Gal successfully onto XyG backbones, indicating their independent mechanism of side chain transfer (Camirand and Maclachlan, 1986).

Figure 1-2

Xyloglucan Structure

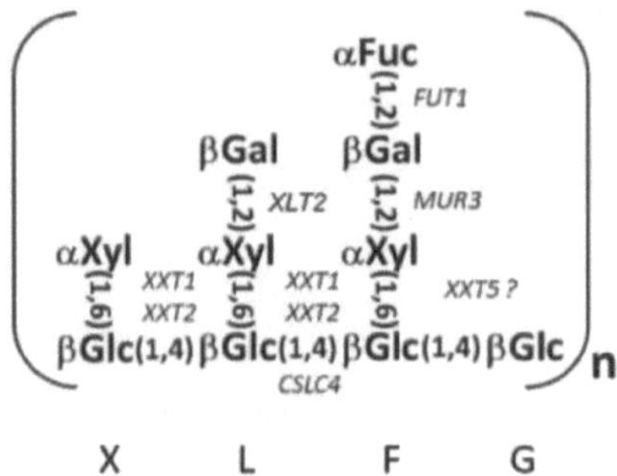

Note. Pattern of linkages for sugars in the XLFG oligosaccharide with the associated glycosyl transferases indicated in upper case letters. Image was obtained from Zabotina (2012). Each of the four common side chain decorations on XyG are described by single letter as follows: 'G' refers to unbranched XyG, whereas X, L, and F indicates a glucosyl backbone substituted with Xyl, Gal-Xyl, and Fuc-Gal-Xyl side chains, respectively.

The cellulose synthase-like C (CslC) family members are involved in xyloglucan backbone synthesis (**Figure 1-3**) and most of the genetic evidence involving the *cslc4* mutant and the quintuple mutant *cslc4 cslc5 cslc6 cslc8 cslc12* in Arabidopsis showed undetectable xyloglucan levels (Kim et al., 2020). Furthermore, xyloglucan xylosyl transferase 1, 2, and 5 (XXT1, 2, 5) catalyze the first step in side chain formation (Culbertson et al., 2018) with undetectable levels of xyloglucan found in *xxt1 xxt2 xxt5* triple mutants (Cavalier et al., 2008). Surprisingly, Arabidopsis plants grew normally in absence of xyloglucan and that appears to run counter to the widely accepted concept that xyloglucan constitutes a load - bearing tether between cellulose fibrils which is important

for plant growth (Kim et al., 2020).

Figure 1-3

Hemicellulose Biosynthesis

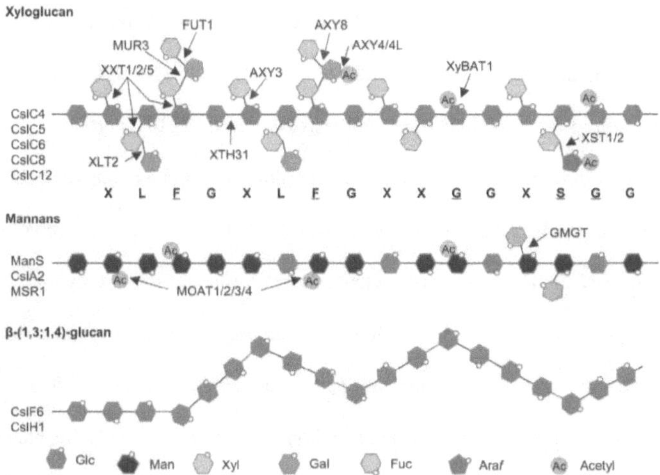

Note. Structural models depicting xyloglucan, mannan and β-1, 3/1, 4- glucan alongside their glycosyltransferases (Zhang et al., 2021).

Xylans. Xylans are the most abundant hemicellulosic polymer with a distinct

pentose-containing β-(1,4)-linked xylosyl backbone present in different plant species

(Smith et al., 2017). Identified xylan synthesizing enzymes are named as IRXs due to the

irregular xylan phenotype found in their *irx* mutants. These enzymes include IRX7,

IRX8, IRX9, IRX9L, IRX10, IRX10L, IRX14, and PARVUS (Wu et al., 2010). Several genes belonging to GT43 and GT47 are implicated in xylan backbone formation. For example, IRX10 and its homolog IRX10L (members of the GT47 family) are β-(1,4)-xylosyl transferases biochemically characterized to be involved in xylan backbone elongation (Urbanowicz et al., 2014). Research supports the cross-species conservation of the role of IRX10 in xylan backbone biosynthesis, given that IRX10 and its homologs from rice and other plant species complemented Arabidopsis *irx10* mutants (Chen et al., 2013). Based on genetic evidence, IRX9, IRX9L, IRX14, and IRX14L (of the GT43 family) are required for xylan backbone synthesis (Wu et al., 2010). Furthermore, mutant analysis of *irx9l* (*spg2*) resulted in a thinner microspore wall and a collapsed pollen phenotype, while *irx9* was indistinguishable from WT (Li et al., 2017). The involvement of GT43 and GT47 family members in xylan biosynthesis suggest that they may assemble to form protein complexes but how they are assembled and organized to form xylan synthase complexes remains unclear (Zeng et al., 2016; Jiang et al., 2016).

Most xylosyl residues of the xylan backbone are substituted with side chains containing either (4-O-methyl) glucuronic acid in dicot plants or arabinose in grasses (Poaceae), which may be further substituted with xylosyl residues or ferulate esters that can be oxidatively cross-linked (Chiniquy et al., 2012). In Arabidopsis, GLUCURONIC ACID SUBSTITUTION OF XYLAN 1−3 (GUX1−3) from the GT8 family add glucuronic acid substitutions onto xylan, which can be further methylated by glucuronoxylan methyltransferases (GXMT1−2; from domain of unknown function family 579) to produce (methyl)glucuronic xylan (GX) (Mortimer et al., 2010). Analysis of *gux1gux2gux3* triple mutants showed undetectable amounts of GlcA on xylan, and resulted in reduced secondary wall thickening and reduction in plant height (Lee et al., 2012). However, unlike xylan biosynthesis in Arabidopsis, xylan elongation and substitution processes are coupled in grass species (Faik, 2010).

Mannans. Mannans are linear glycan chains of β-(1,4)-linked mannose residues, while glucomannan refers to mannan interspersed with β-(1,4)-linked glucose. Both mannans and glucomanannans can be further modified with galactose residues linked to mannosyl residues at the O-6 site. Mannans are also storage forms of polysaccharides. Indeed, an investigation on how polysaccharides are stored in the endosperm of guar (*Cyamopsis tetragonoloba*) seeds led to the discovery of mannan synthase ManS (a CslA protein) that was identified as a galactomannan synthase (Dhugga et al., 2004). Following this discovery, more CSL proteins were characterized and found to be required for β-glycan formation (Burton et al., 2006; Kim et al., 2020). Similarly, mannan O-acetyltransferases (MOATs) modify the mannosyl residues of the mannan backbone with acetyl epitopes (**Figure 1-3**) (Zhong et al., 2018).

β-1, 3/1, 4-Mixed-linked Glucans (MLGs). MLGs are characterized by unbranched β-(1,4) -linked glucans interspersed with β-(1,3) -glucosyl linkages (**Figure 1-3**). MLGs in Poales are developmentally regulated and play crucial roles in cell expansion and cellulose cross-linking (Scheller and Ulvskov, 2011). Although CslF and its homologs are absent in Arabidopsis, expression of rice CslF6 in Arabidopsis produced β-(1,3;1,4)-glucans in Arabidopsis tissues (Vega-Sanchez et al., 2012). Furthermore, the ability of CslF6 to target the plasma membrane indicated that the β-(1,3;1,4)-glucan production likely occurs at sites other than the Golgi apparatus (Wilson et al., 2015). In addition, CslH family members have been reported to have the ability to synthesize β-(1,3;1,4)-glucan (Doblin et al., 2009).

Pectins

Pectins are a group of cell wall polysaccharides with α-(1,4)-linked galacturonic acid (GalA) in their backbone that is further modified to form different configurations such as unbranched homogalacturonan (HG), rhamnogalacturonan I (RG-I), and complex branched rhamnogalacturonan II (RG-II) and Xylogalacturonan (**Figure 1-4**) (Atmodjo et al., 2013).

Figure 1-4

Structure of Different Components of Pectin Polysaccharides in Plants

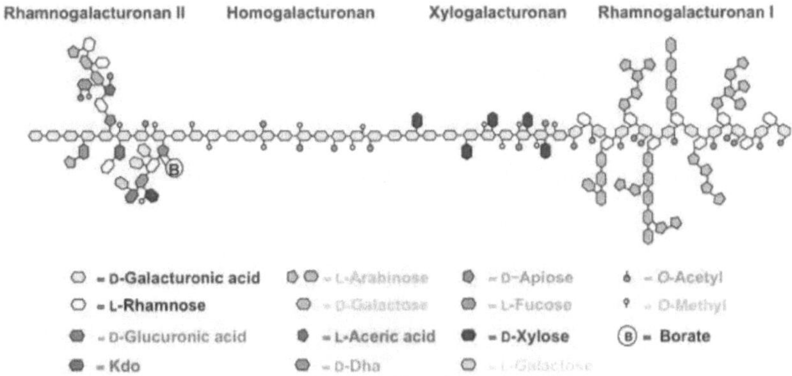

Note: The four main types of pectin are homogalacturonan, rhamnogalacturonan I (RG-I), xylogalacturonan, and rhamnogalacturonan II (RG-II). Homogalacturonan (HG) is comprised mainly unbranched α-1,4-galacturonic acid (GalA) chains; xylogalacturonan differs slightly from HG by the presence of xylose side chains. The RG-I backbone is made up of alternating units of rhamanosyl (Rha) and GalA units, while RG-II, which is the most complex pectin polymer, contains an α-1,4-GalA backbone that is extensively decorated by other sugar molecules such as aceric acid, 3-deoxy-D-manno-2-octulosonic acid (KDO), and 3-deoxy-D-lyxo-2-heptulosaric acid (DHA) and also crosslinked with other polymers by boron (Harholt et al., 2010).

Homogalacturonans. Homogalacturonan (HG) is known for its structural simplicity and abundance in most types of plant cell walls (Sorensen et al., 2011). Structurally, it is composed of a linear chain of α-1,4-linked galacturonic acid (GalA) residues that can either be methyl-esterified at the C6 carboxyl group or acetylated at the *O*2 or *O*3 position (Yapo et al., 2007). HG containing de-methylesterified GalA can form intermolecular cross-links that are coordinated by calcium ions (Ca^{2+}), resulting in an "egg-box" conformation (Braccini and Perez, 2001). Further HG side chain modifications involving xylose (forming xylogalacturonan) linked to the O3 position, or apiose (forming apiogalacturonan) linked to the O2 or O3 position have been reported (Yapo et al., 2007).

The GALACTURONOSYLTRANSFERASE (GAUT) family consists of enzymes thought to be involved in HG biosynthesis and reside in GT8. Such GT8 family members elongated GalA chains using UDP-GalA as a substrate (Atmodjo et al., 2013). *GAUT* and *GAUT-LIKE (GATL)* genes are present in most plant species (McCarthy et al., 2014), and some of these genes are functionally relevant in HG synthesis (Kong et al., 2011). For example, Arabidopsis GAUT1 was the first GAUT that was biochemically characterized to demonstrate galacturonosyltransferase (GalAT) activity, and additional GAUTs have yet to be characterized (Voiniciuc et al., 2018). Moreover, some GAUTs, are thought to act as scaffolding in protein complexes involving GAUTs (Atmodjo et al., 2011).

HG are synthesized in a highly methyl-esterified form in the Golgi through the actions of HG methyltransferases that use S-adenosylmethionine (SAM) as a substrate. One such putative HG methyltransferase is QUASIMODO2 (QUA2), which showed show aberrant HG extractability and cell adhesion defects in *qua2* mutants (Krupkova et al., 2007). During suppressor screens for the cell adhesion defect of *qua2* mutants, ESMERALDA1, which encodes a putative *O*-fucosyltransferase was discovered (Verger et al., 2016). Two other proteins, COTTON GOLGI RELATED 2 (CGR2) and CGR3, were reported as putative pectin methyltransferases, and microsomal membranes isolated from CGR overexpressor lines demonstrated enhanced HG methyltransferase activity (Kim et al., 2015). Despite this discovery, it is unclear whether QUA2, CGR2, and CGR3 work coordinately or distinctively during HG polymerization in the Golgi (Kim et al., 2015).

Rhamnogalacturonan-I (RG-I). RG-I makes up approximately 20-35% of pectic polysaccharides. The RG-I backbone is comprised of alternating GalA and rhamnose (Rha) residues that are linked by α-1,2 and α-1,4 bonds, respectively, which is further modified or decorated with arabinans, galactans, and arabinogalactans (Yapo, 2011). In some cases, galactan side chains on RG-I contain a terminal fucose (Nakamura et al., 2001), while both arabinan and galactan side chains can be modified with terminal ferulic acid moieties (Ishii, 1997), that can potentially cross-link to generate RG-I dimers in the cell wall (Ralet et al., 2005). Several enzymes such as rhamnosyltransferase enzymes demonstrated their involvement in the RG-I backbone formation (Takenaka et al., 2018). Similarly, GALACTAN SYNTASE (GALS) and the arabinosyltransferases of the ARABINAN DEFICIENT (ARAD) family enzymes are thought to be involved in the synthesis of galactan and arabinan side chains of RG-I respectively (Ebert et al., 2018).

Rhamnogalacturonan-II (RG-II). Among the biomolecules that make up pectins, RG-II is the most complex, with some structural variations among vascular plants (Pabst et al., 2013). About 13 different types of sugars have been discovered in RG-II that may be linked by up to 22 unique glycosidic bonds to form 6 different unique side chains (Ndeh et al., 2017). RG-II molecules can be cross-linked via borate diesters that interconnect their apiose residues (Ishii and Matsunaga, 1996), and structural variations in their crosslinking patterns exist across developmental stages and in some taxa, despite their conservation across almost plant species (O'Neill et al. 2004).

A large suite of catalytic activities is required to construct the complex side chains of RG-II (Mohnen, 2008). Despite the large suite of enzymes involved in the synthesis of RG-II, only a few of the enzymes required have been identified, including a glucuronosyltransferase (Iwai et al., 2002) and two xylosyltransferases (Egelund et al., 2006).

Lignin

Lignin is a complex aromatic biopolymer based on 4-hydroxyphenylpropanoids (Simmons et al., 2010). Three major lignin monomers are hydroxyphenyl units (H units), guaiacyl units (G units), and syringyl units (S units), which are predominantly interconnected through ether linkages and carbon-carbon linkages (Albersheim et al., 2010). Literature reports the formation of lignin-carbohydrate complexes between lignin and carbohydrate moieties (mainly hemicellulose) which plays a significant role in establishing wood structure (Du et al., 2014)

Plant Cell Wall Proteins

Plant cell wall proteins are glycosylated and mainly consist of structural and enzymatic proteins (Josè-Estanyol and Puigdomènech, 2000). Plant cell wall proteins (CWPs) and peptides are important players in wall assembly and remodeling during plant growth and development. Among cell wall proteins (CWPs), the hydroxyproline (Hyp)-rich *O*-glycoproteins (HRGPs) are complex macromolecules with various structures and functions. They exist as moderately glycosylated extensins (EXTs), hyperglycosylated arabinogalactan-proteins (AGPs) and Hyp/Pro-rich proteins (H/PRPs) that may be non-, weakly-or highly glycosylated. Similarly, hybrid HRGPs, composed of HRGP modules from different families, and chimeric HRGPs, composed of one or more HRGP modules within a non-HRGP protein, have been identified previously and considered part of the HRGP superfamily (Showalter et al., 2010). Each HRGP sub-family is characterized by repetitive consensus sequences which determine the way they are glycosylated (Kieliszewski et al., 1994; Estevez et al., 2006).

Cell wall glycoproteins can be N-glycosylated, which requires a consensus sequence of Asn-X-Ser/Thr (where X can be any amino acid except proline). Notably, not all protein sequences with this motif are glycosylated, which may be a reflection of the absence of the enzymatic machinery responsible for carbohydrate assembly or the masking of the consensus motif sequence during protein folding which makes it not susceptible to modification (Han et al., 1992). The hallmark of N-glycosylation in all eukaryotes involves the *en bloc* transfer of the pre-assembled $Glc_3Man_9GlcNAc_2$ oligosaccharide from the lipid carrier dolichol pyrophosphate to selected asparagine residues in the Asn-X-Ser/Thr consensus sequence within nascent polypeptides (Strasser, 2016). Plant cell wall glycoproteins such as AGPs are mainly O-glycosylated mainly through their hydroxyproline residues while O-glycosylation of serine with galactose and hydroxyproline with arabinose have been described for EXTs and requires a Ser-Hyp$_4$ environment (Kieliszewski et al., 1994). In O-glycosylation, all the sugars are added successively and in a coordinated fashion by the transfer from their nucleotide derivates and by the activity of specific GTs in the Golgi apparatus

(Kieliszewski et al., 1994). About 166 HRGPs genes were identified in Arabidopsis including 89 AGPs, 59 EXTs, and 18 PRPs by using the tools of bioinformatics (Showalter et al., 2010).

Extensins (EXT). EXTs are cell wall glycoproteins, whose sequence contains a signal peptide allowing them to be exported to the wall (Josè-Estanyol and Puigdomènech, 2000). They are characterized mainly by the repeating pentapeptide sequence Ser-Hyp- Hyp-Hyp-Hyp (SO_4) but structural variations such as SO_3 and SO_5 as well as monogalactosylation of some Ser residues and oligoarabinosylation of most Hyp residues do exist (Showalter, 1993). EXTs are rod-like glycoproteins that exist in a polyproline II helix stabilized by the oligoarabinosides wrapping around the protein backbone (Shpak et al., 2001). Similarly, some EXTs contain Tyr residues that can form intramolecular isodityrosine crosslinks that cause "kinks" in the rods, or intermolecular di- isodityrosine or pulcherosine crosslinks that result in the formation of EXT networks in the wall (Brady et al., 1998). Such crosslinks have been reported to regulate plant growth and form a defensive barrier to prevent pathogen infection (Qi et al., 1995).

EXT biosynthesis involves several posttranslational modifications that include signal peptide processing in the endoplasmic reticulum (ER) (Velasquez et al., 2012), hydroxylation of peptidyl-proline residues into hydroxyproline (Hyp) residues in the ER-Golgi apparatus (Yuasa et al., 2005), O-glycosylation on Hyp and Ser by the action of several glycosyltranferases, mostly in the Golgi apparatus and finally, covalent intra- and inter-molecular crosslinking at the plant cell wall level (Held et al., 2004).

Extensin Hydroxylation. Plant prolyl 4-hydroxylases (P4Hs) are membrane-bound enzymes that catalyze the hydroxylation of peptidyl-proline to yield hydroxyproline by introducing an oxygen atom between the hydrogen and carbon atoms of the peptidyl-proline (Lamport, 1963). P4Hs belong to a family of 2-oxoglutarate-dependent dioxygenases that require 2-oxoglutarate and O_2 as co-substrates, Fe^{2+} as a cofactor and ascorbate to prevent the rapid inactivation by self-oxidation (Koski et al., 2009). The activities of P4Hs in the secretory pathway provides reactive hydroxyl groups that initiate O-glycosylation processes during EXT biosynthesis. Several *in vitro* and *in vivo* characterizations of plant P4Hs have been performed in *Arabidopsis thaliana* (Asif et al., 2009), *Nicotiana tabacum* (Yuasa et al., 2005), *Dianthus caryophyllus* (Vlad et al., 2007), and the green alga *Chlamydomonas reinhardtii* (Keskiaho et al., 2007). Although, 13 P4Hs have been identified in the Arabidopsis genome, only the activity of AtP4H1 and AtP4H2, have been fully characterized *in vitro* (Koski et al., 2009). Notably, AtP4Hs hydroxylate proline-rich peptides of both arabinogalactan-proteins (AGPs) and EXTs.

Arabidopsis T-DNA insertional mutants (*Atp4h2,5,13*) exhibited a short root hair phenotype, while the overexpression of these P4Hs displayed extra-long root hairs (Velasquez et al., 2011), indicating that they are essential for polarized cell expansion in root hairs. In a related experiment involving *Chlamydomonas reinhardtii*, the suppression of the gene for 1 of the 10 *Chlamydomonas reinhardtii* prolyl 4-hydroxylases (CrP4H1), which hydroxylates algal HRGP sequences, led to a defective cell wall (Keskiaho et al., 2007), suggesting a major unique structural role of certain HRGPs as targets of CrP4H1 in this unicellular green alga. Also, a potential role for P4Hs in hypoxia stress was suggested along with their potential use as oxygen sensors (Asif et al., 2009).

Extensin Glycosyltransferases. Several genes/proteins involved in EXT glycosylation have been identified and characterized. Specifically, at least nine genes have been discovered in the process (**Figure 1-5**). One of these genes, SERINE GALACTOSYLTRANSFERASE 1 (SGT1), belonging to the GT96 family, was identified in *Chlamydomonas reinhardtii* by sequencing a protein with O- α-serine galactosyltransferase activity with respect to the addition of single galactose residues to Ser residues in SO_4 sequences (Saito et al., 2014). The other eight genes are involved in EXT glycosylation encode arabinosyltransferases that preferentially transfer arabinose residues to specific substrates on the EXT backbone. Hydroxyproline O-β-arabinosyltransferases (HPATs), HPAT1, HPAT2, and HPAT3 were found to add single arabinose residues to Hyp residues in SO_4 sequences (Ogawa-Ohnishi et al., 2013). These genes were identified in Arabidopsis following affinity purification and sequencing proteins that demonstrated HPAT activity. REDUCED RESIDUAL ARABINOSE 1-3 (RRA1, RRA2, and RRA3), belonging to the GT77 family, were reported to be involved in adding the second arabinose residue in a β-1,2 linkage (Velasquez et al., 2011). Similarly, XYLOGLUCANASE 113 (XEG113), also belonging to the GT77 family, was found to encode an arabinosyltransferase which adds the third arabinose residue also in a β-1,2 linkage (Gille et al., 2009). In addition, EXTENSIN ARABINOSE DEFICIENT (ExAD), a member of the GT47 family, demonstrated the ability to add the fourth arabinose in an α-1,3 linkage (Velasquez et al., 2012).

Figure 1-5

EXT Biosynthesis

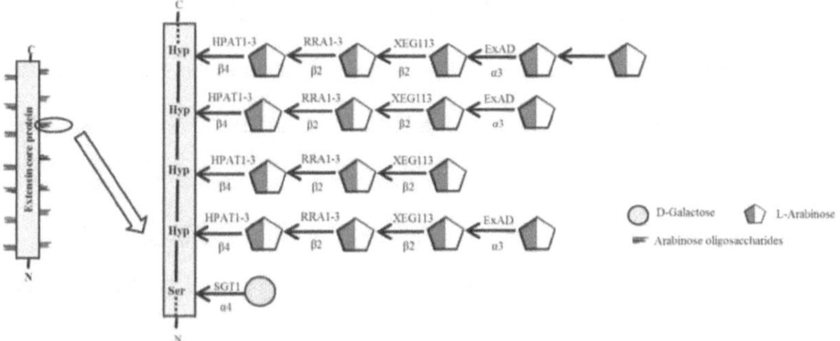

Note. Sites of action of known glycosyltransferases involved in EXT biosynthesis are depicted within a representative EXT glycomodule sequence found in an EXT molecule (Showalter and Basu, 2016).

Extensin Function. Most of the functions ascribed to EXTs have been mainly

deduced through genetic evidence during mutant analysis. This is a powerful approach to

determine the functional contributions of the carbohydrate moieties that decorate EXTs

and other plant cell wall proteins. For example, *sgt1* mutants displayed larger rosettes and

longer roots (Velasquez et al., 2015). Similarly, *hpat* mutants demonstrated several

pleiotropic phenotypes including impaired pollen tube growth, early senescence, early

flowering, defects in cell wall thickening, enhanced hypocotyl elongation, and shorter

root hairs (Ogawa-Ohnishi et al., 2013). In a related study, HPAT1, HPAT2 and HPAT3

demonstrated that they are essential in male gametophyte transmission in Arabidopsis, as

hpat1 hpat3 and *hpat1 hpat2 hpat3* exhibited shorter silique lengths, less silique numbers, as well as reduced pollen tube lengths (MacAlister et al., 2016). Also, *rra* and *xeg113* mutants displayed reduced root hair growth (Gille et al., 2009).

Moreover, genetic knockouts of EXT protein backbones have displayed mutant phenotypes that elucidated their biological roles in plant growth. For example, an extensin mutant, *root-, shoot-, hypocotyl-defective (rsh) ext3*, has been shown to have a nearly lethal phenotype (Cannon et al., 2008). Similarly, *leucine-rich repeat extensins* 1 and 2 (*lrx1* and *lrx2*) showed aberrant root hairs (Ringli, 2010), while a *proline-rich extensin-like receptor kinase (perk13)* mutant displayed longer root hairs (Humphrey et al., 2007)

Arabinogalactan-Proteins. Arabinogalactan-proteins (AGPs) are plant glycoproteins that consist of a core-protein backbone *O*-glycosylated by complex carbohydrates with galactose and arabinose serving as main components, while rhamnose, glucuronic acid and other monosaccharide residues are present as minor components (Seifert and Roberts, 2007). AGPs consist of a core protein of varying lengths modified with arabinogalactan side chains; about half of all AGPs also contain C-terminal GPI lipid anchors which allow attachment to the outer portion of the plasma membrane (Schultz et al., 2000).

AGP Core Proteins. AGPs are characterized by an N-terminal cleavable signal sequence and, in many cases, a hydrophobic C-terminal domain that is post translationally modified by a GPI lipid anchor (Seifert and Roberts, 2007). Some mature AGP backbones are peptides that are only 10 to 13 residues long and are termed AG peptides. Both classical AGPs and AG peptides consist of a single central domain rich in

Pro, Ala, Ser, and Thr (Schultz et al., 2000). Some AGP species contain a short lysine-rich domain between the proline-rich domain and the hydrophobic C terminus (Schultz et al., 2002), whereas a much larger group of AGPs known as fasciclin-like AGPs (FLA) are comprised of AGP-like glycosylated regions and putative cell adhesion domains known as fasciclin domains (Johnson et al., 2003). The occurrence of different types of hydroxyproline (Hyp)-containing repeats in cell wall glycoproteins led to the Hyp-contiguity hypothesis, which predicts arabinosylation of contiguous Hyp residues and arabinogalactosylation of clustered, noncontiguous Hyp residues in extensin-like or AGP-like glycomodules, respectively (Shpak et al., 2001). The repetitive consensus motifs are thought to guide the post translational attachment of specific *O*-glycans and can be referred to as either AGP or EXT glycomodules. Further experiments towards verifying the Hyp contiguity hypothesis found that *O*-glycans were attached to glycomodules on artificial AGPs and EXTs in plants (Shpak et al., 1999) and indicated that contiguous Hyp residues are modified by short, linear oligoarabinosides and isolated Hyp residues are galactosylated and subsequently modified by type II AG (Kieliszewski, 2001). Although many known AGPs contain XP dipeptide repeats (where X stands for A, S, or T), variants of this repeat have been found in other AGPs (Shi et al., 2003; van Hengel et al., 2003).

 Hydroxylation of AGPs. Before the initiation of *O*-glycosylation, the backbone peptides are post-translationally modified by prolyl hydroxylation, which requires the coordinated actions of proly-4-hydroxylases (P4H), oxidoreductases that depend on Fe^{2+}, 2-oxoglutarate, O2, and ascorbate (Seifert and Roberts, 2007). Both *AtP4H-1* and *AtP4H-*

2 genes encode active P4H, but with different substrate specificities (Tiainen et al., 2005). In the Arabidopsis genome, 13 putative P4H genes have been discovered (Vlad et al., 2007), and their prolyl hydroxylation activities and efficiency are influenced by the peptide sequence, as synthetic $(SP)_n$ and $(AP)_n$ glycomodules are completely hydroxylated; whereas, $(TP)_n$ and $(VP)_n$ glycomodules are only partially hydroxylated (Tan et al., 2003).

Structure of AGP Glycan. AGPs are *O*-glycosylated at one or more hydroxyproline residues on AGP protein cores by AG type II polysaccharides **(Figure 1-6)**. Generally, *O*-glycosylated AGPs consist of primarily (1→3)-β-galactan and (1→6)-β-linked galactan chains connected to each other by (1→3, 1→6)-linked branch points, along with side chain modifications by glucuronic acid (GlcA), fucose (fuc), rhamnose (Rha), xylose (Xyl), and other sugars (Seifert and Roberts, 2007). Individual type II AG structures vary between species and between tissues in the same species, and the heterogeneity of the Type II AG glycan likely provides contact surfaces for interaction with other matrix molecules (Seifert and Roberts, 2007). Although, the type II glycan structure is similar to other AGP glycans, some differences exist as observed in the *Artemisia vulgaris* pollen allergen that is *O*-glycosylated at hydroxyproline residues with a (1→6)-β-galactan core and decorated by 5-; 2,5-; 3,5-; and 2,3,5-linked arabinan side chains (Leonard et al., 2005) and also in *Physcomitrella patens* AGP where the (1→3)-β-galactan backbone appears to contain linear (1→5)-α-arabinan side chains (Lee et al., 2005).

Figure 1-6

Type II Arabinogalactan Polysaccharide Attached to a Hydroxyproline Reside on an

AGP

Note. The type II arabinogalactan depicted above was elucidated from synthetic GFP:(AP)51, expressed in tobacco BY2 cells (Tan et al., 2004).

The glycan moiety of AGPs accounts for more than 90% sugars and the remainder containing the AGP protein core. Synthetic phenyl-glycoside dyes (β-Yariv reagents) can specifically bind to the β-1,3-galactan moiety of AGPs (Kitazawa et al., 2013). Similarly, various monoclonal antibodies that recognize different AGP glycan epitopes (Seifert and Roberts, 2007). Both Yariv reagent and AGP-specific antibodies are useful tools to elucidate the location of certain AGP glycan epitopes in plant tissues.

AGP Glycosyltransferases. GTs are enzymes involved in the transfer of a sugar moiety from an active sugar donor to an active sugar acceptor, forming a glycosidic bond (Cantarel et al., 2009). An attempt to understand the mechanisms involved in AGP biosynthesis and modification has led to the identification and characterization of several classes of enzymes that may prove useful in the genetic manipulation of plant cell wall genes for improved biomass. These enzymes belong to the CAZy superfamily and several GTs and glycosyl hydrolases involved in AGP biosynthesis have been discovered.

At least eighteen different GT genes are reported to be involved in AGP glycosylation and this include genes encoding β-galactosyltransferases (GALTs), α-arabinosyltransferases, fucosyltransferases (FUTs), rhamnosyltransferases, and glucuronosyltransferases (GLCATs) (Showalter and Basu, 2016). AGP glycosylation begins with the action of hydroxyproline *O*-β-galactosyltransferases, responsible for the addition of the first galactose residue onto Hyp residues in AGP core proteins (Showalter and Basu, 2016). Eight genes encode this activity and were identified using bioinformatics and verified by heterologous expression and protein sequencing of products obtained during affinity chromatography coupled with enzyme assays and by genetic mutant analysis (Basu et al., 2015; Ogawa-Ohnishi and Matsubayashi, 2015). These genes, named *GALT2*, *GALT3*, *GALT4*, *GALT5* and *GALT6*, encode a GALT domain as well as a GALECTIN domain, while *HPGT1*, *HPGT2*, and *HPGT3* encodes a GALT domain but lacks a GALECTIN domain. All these genes belong to the GT31 family. GALT1, also encodes both of these domains, but is not involved in AGP glycosylation; instead, it adds galactose to Lewis a structures (Strasser et al., 2007). The subcellular location of GALT3-6 and HPGT1 is restricted to the Golgi, but GALT2 is localized in both the ER and Golgi and indicates that AGP glycosylation may be initiated in the ER before transiting to the Golgi (Basu et al., 2013).

Arabidopsis GALT2 was first identified and biochemically characterized to be involved in catalyzing the addition of the first galactose residue on hydroxyproline residues on synthetic AGP peptides (Basu et al., 2013). Similarly, *At1g32930*, another GT31 gene (*AtGALT31A*), was characterized by using a recombinant *AtGALT31A* gene expressed in *Escherichia coli* and *Nicotiana benthamiana*. This demonstrated β-1,6-GalT activity, as exemplified by the elongation of β–1,6- galactan side chains of AGP glycans Geshi et al., 2013). KNS4/UPEX1 belonging to GT31 family demonstrated β-1,3-galactosyltransferase activity, but its activities may involve both AGPs and RG-I (Suzuki et al., 2017). GALT29A, demonstrated the same enzymatic activity as GALT31A, with a demonstrated ability to add β–1,6-galactose to β–1,3-galactans of AGP glycans (Dilokpimol et al., 2014). GALT29A was identified as being co-expressed with GALT31A, and protein-protein interaction assay by fluorescence resonance energy transfer analysis, verified GALT29A and GALT31A proteins do interact and exist in an enzyme complex (Dilokpimol et al., 2014). It was suggested that this may be an important molecular mechanism regulating the synthesis of β-1,6-galactan side chains of type II AGs during plant development (Knoch et al., 2014).

Out of eleven β-glucuronosyltransferase (GLCAT) genes in the GT14 family, five Arabidopsis GLCAT genes (*GLCAT14A-E*) have been identified and biochemically characterized to be involved in the glucuronidation of AGP glycans (Knoch et al., 2013; Dilokpimol and Geshi, 2014; Lopez-Hernandez et al., 2020). Arabidopsis *GLCAT14A* (*At5g39990*) gene was first expressed in *Pichia pastoris* and demonstrated β-GLCAT activity by adding GlcA to both β-1,6- and β-1,3-galactans. Similarly, *At5g15050* and

At2g37585 also demonstrated the same β-GLCAT activities and were named *AtGLCAT14B* and *AtGLCAT14C*, respectively (Dilokpimol and Geshi, 2014). Interestingly, GlcA has the ability to bind and release calcium in a pH dependent manner, and this is mediated by the paired carboxylic ends of GlcA which is negatively charged and provides a potential site for calcium binding (Lamport and Varnai, 2013; Lopez-Hernandez et al., 2020). In type II AGs from *Arabidopsis*, most of the GlcA substituted is 4-*O*-methylglucuronosyl (4-*O*-Me-GlcA; Tryfona et al., 2012). A recent study identified two Golgi-localized enzymes, arabinogalactan methyltransferase 1 (AGM1) and 2 (AGM2), belonging to DUF579 family, that demonstrated GlcA-*O*-methylation activity on AGPs (Temple et al., 2019). While the *agm1* mutant showed reduced methylation of GlcA on root AGPs, AG GlcA methylation was absent in the *agm1agm2* double mutant (Temple et al., 2019). AtFUT4 and AtFUT6, encoded by *At2g15390* and *At1g14080*, respectively, were reported to fucosylate AGPs and were identified as α-1,2-fucosyltransferases involved in the biosynthesis of AGP glycans when heterologously expressed in BY2 cells (Wu et al., 2010). Another gene/enzyme belonging to GT77 family, *REDUCED ARABINOSE YARIV 1*, or *RAY1*, encodes an arabinosyltransferase and demonstrated β–arabinosyltransferase activity to methyl β–galactose when heterologously expressed in *Nicotiana benthamiana* (Gille et al., 2013). However, RAY1's function in AGP biosynthesis is debatable given that α-linked arabinose, and not β-linked arabinose, is reported in AGPs (Knoch et al., 2014).

AGP and AGP Glycosyltransferase Mutants. The important role AGP glycosylation plays in the overall AGP function in plant growth has been uncovered mostly through genetic mutant analysis. The *galt2-6* single mutants showed some mutant phenotypes such as reduced root hair length and density for *galt2*, *galt3*, and *galt5*, reduced seed set for *galt4* and *galt6*, reduced adherent seed coat mucilage for *galt3* and *galt6*, and premature senescence for *galt6* (Basu et al., 2013). However, severe physiological defects were observed in higher order *galt* mutants which exemplified the genetic redundancy in their physiological functions. For example, *galt2galt5* double mutants displayed a more severe and pleiotropic mutant phenotypes with respect to root hair length and density and seed coat mucilage, delayed flowering, and shorter siliques (Basu et al., 2013). These findings were further corroborated in a similar study involving *hpgt* mutants which showed no obvious phenotypes under normal growth conditions in *hpgt1-3* single mutants, but severe defects in physiological functions were observed in *hpgt1hpgt2hpgt3* triple mutants, which had longer lateral roots, increased root hair length and density, thicker roots, smaller rosette leaves, shorter petioles, shorter inflorescence stems, reduced fertility, and shorter siliques (Ogawa-Ohnishi and Matsubayashi, 2015). Loss of function of KNS4/UPEX1 demonstrated that it is required for pollen exine development (Suzuki et al., 2017), while the genetic knock-out of GALT31A resulted in embryo lethality that was due to the abnormal cell division in the hypophysis (Geshi et al., 2013). The *galt2 galt3 galt4 galt5 gal6* quintuple mutants generated by CRISPR-Cas9 approach showed defects in both vegetative and reproductive tissues, and were characterized by impaired root growth, abnormal pollen development, shorter siliques, and reduced seed set (Zhang et al., 2021). The *ray1* mutant displayed abnormal growth that may be reflective of reduced arabinose levels in type II AGs (Gille et al., 2013);

however, the biochemical role of RAY1 as previously mentioned remains elusive (Showalter and Basu, 2016). Similarly, loss of function of FUT4 and FUT6 displayed no obvious phenotypes except for the root growth inhibition that was reported for *fut4fut6* mutants when grown in salt (Liang et al., 2013; Tryfona et al., 2014). This finding underscores the importance of AGP fucosylation in salt stress. Several type II AG specific GLCATs have demonstrated their importance in plant growth processes. For example, previous work showed that the genetic knock-out of *GLCAT14A* showed enhanced cell elongation (Knoch et al., 2013), a finding that was later challenged by a recent work (Lopez-Hernandez et al., 2020). Similarly, phenotypic analyses found that the *glcat14a glcat14b* and *glcat14a glcat14b glcat14c* mutants displayed significant delays in seed germination, reductions in root hair length, reductions in trichome branching, and accumulation of defective pollen grains. Additionally, both *glcat14b glcat14c* and *glcat14a glcat14b glcat14c* displayed significantly shorter siliques and reduced seed set (Zhang et al., 2020). Also, *agm1agm2* double mutants did not display a growth or fertility phenotype, but an earlier study demonstrated that the methyl group on GlcA is essential for the effectiveness of a signaling molecule in pollen tube guidance (Mizukami et al., 2016).

Literature exists that suggest that the genetic disruption of AGP protein cores impact normal physiological functions especially in reproductive tissues. Mutant characterization of *agp6*, *agp11*, and *agp6 agp11* mutants showed defects in pollen germination, reduced pollen tube elongation and pollen release (Coimbra et al., 2009). Similarly, TTS (Transmitting Tissue-Specific) observed to be a 120kD AGP, displayed reduced female fertility and less fruit set phenotypes in TTS RNAi knock-downs (Wu et

al., 1995). AGP4, also known as JAGGER, was implicated in ensuring polytubey block

in Arabidopsis, as *jagger* mutants displayed an increased attraction of supernumerary

pollen tubes to an ovary (Pereira et al., 2016). Similarly, a terminal disaccharide sugar, β-

methyl-glucuronosyl galactose (4-Me-GlcA-β-1,6-Gal) of AGPs were discovered to

induce pollen tube competency to respond to ovular guidance *in vitro* (Mizukami et al.,

2016).

 Interactions of AGPs With Other Biological Macromolecules During Cell Wall

Biosynthesis. Deconstructing plant cell walls center on understanding the regulatory

mechanisms that control how plant cell walls interact with other molecules. Some AGPs

have been demonstrated to interact with other cell wall polymers via their *O*-glycans and

their close proximity with other plasma membrane proteins such as receptor-like kinases

(RLKs) and other cell wall polysaccharides (Dilokpimol et al., 2014). The

APAP1/AGP57C protein was first discovered as a minor AGP fraction secreted by

suspension cells and the *apap1* knock out mutants in *A. thaliana* showed substantial

alterations in some cell wall polysaccharides due to type II AG crosslinking with other

cell wall polysaccharides (Tan et al., 2013). Similarly, some AGP glycans have been

observed to act in the same genetic pathway with their interacting partners through

genetic studies. Both the *galt2 galt5* double mutant and the *hpgt1hpgt2 hpgt3* triple

mutant displayed a root and seed coat mucilage phenotype similar to the salt overly

sensitive 5 (*sos5-1*), a mutant in the FASCICLIN LIKE ARABINOGALACTAN-

PROTEIN 4 (FLA4) locus (Basu et al., 3013). The *sos5-1* mutant was identified in a mutant screen for salt oversensitive root growth (Shi et al., 2003) and found to act in a linear genetic pathway with two redundantly acting leucine-rich receptor kinases (LRR-RLK) named FEI1 and FEI2 (Xu et al., 2008). Similarly, the *sos5-1* mutant displayed seed coat mucilage adhesion defect that resulted in aberrant mucilage pectin structure (Griffiths et al., 2016). Further analysis of *galt2 galt5 sos5 fei1 fei2* quintuple mutants showed that GALT2 and GALT5 act non-additively with both SOS5, FEI1 and FEI2. This genetic evidence suggests that the glycosylation of AGPs is required for the genetic function of SOS5 and FEI1 FEI2 (Basu et al., 2016).

Subcellular Coordination During Plant Cell Wall Assembly

Although the majority of plant cell wall synthesis occur mainly in the Golgi and at the plasma membrane, the endoplasmic reticulum (ER) has been found to be involved in ensuring proper protein folding and assembly, protein glycosylation initiation, and protein transport to the Golgi (Hoffman et al., 2021). *N*-glycosylation is initiated in the ER and is an important post-translational modification event that affects folding, trafficking/retention and enzymatic activity of many cell wall proteins (Zeng et al., 2018). Many cell wall proteins, e.g., arabinogalactan-proteins, are highly *O*-glycosylated at hydroxyproline residues (Seifert, 2020), and enzymes involved in the first committed step during AGP biosynthesis are localized to the ER (Velasquez et al., 2011; Basu et al., 2015). Apart from cell wall synthesis in the Golgi and at the plasma membrane, lignin monomers for secondary cell wall synthesis are made by soluble cytoplasmic enzymes and ER-associated cytochrome P450s and are then exported to the cell wall (Gou et al., 2018).

Plant cells contain numerous, independent, and highly mobile Golgi stacks that play a critical role in the production of cell wall matrix polysaccharides, modification of protein glycosylation, and assembly/trafficking of CSCs (Hoffman et al., 2021). Each Golgi stack contains a full complement of resident proteins and enzymes that work cooperatively to contribute to cell wall matrix polysaccharide and glycoprotein biosynthesis (Meents et al., 2019). Nucleotide sugar transporters are also involved in these glycosylation events, as they translocate activated sugar precursors for polysaccharide biosynthesis from the cytosol to the Golgi lumen (Rautengarten et al., 2014), and type-I and type-II GTs use these substrates to produce and/or modify the polysaccharide main and side chains, respectively, in the Golgi lumen (Anderson, 2016). Regarding the coordinated activities of GTs that drives cell wall synthesis, the sequential model and the enzyme complex-mediated model have been proposed, although it is currently unclear which prevailing model is being exploited by different Golgi cisternae (Gimeno-Ferrer et al., 2017). The sequential model predicts that different GTs occupy distinct Golgi sub-compartments, and polysaccharide synthesis is coordinated in a stepwise manner in different cisternae by different enzymatic reactions. For example, modification of the xyloglucan backbone was localized in medial cisternae while xyloglucan side chain modifications were localized in trans-cisternae (Parsons et al., 2019). The complex-mediated model predicts that GTs interact together in one multi-enzyme complex to catalyze glycosylation events in the Golgi (Hoffman et al., 2021). This model was demonstrated in pectin homogalacturonan GTs, GAUT1 and GAUT7, whose interaction is critical for their retention in the Golgi (Atmodjo et al., 2011).

Similarly, hemicellulose GT targeting also appears to be dependent upon complex formation, as simultaneous co-expression of the xylan biosynthetic genes *IRX9*, *IRX10*, and *IRX14A* from asparagus was required for enzyme localization to the Golgi (Zeng et al., 2016; Jiang et al., 2016), while bimolecular fluorescence complementation (BiFC) and pull-down assays provided further evidence that xyloglucan backbone glucan synthases (CSLCs), xylosyltransferases (XXTs), galactosyltransferases (MUR3 and XLT2), and fucosyltransferases (FUTs) can physically associate *in vivo* (Lund et al., 2015). Following maturation in the trans-most Golgi cisterna, cell wall matrix polysaccharides, proteins, and CSCs move to the trans Golgi network (TGN) to be sorted into secretory vesicles (Kang et al., 2011). Plant TGNs are engaged both in receiving and sorting cargo from both the secretory and endocytic pathways, and therefore, they are often referred to as TGNs/early endosomes (EEs) (Viotti et al., 2010). TGNs/EEs belong to structurally and biochemically distinct subdomains, each of which play distinct but overlapping roles during cell wall synthesis (Uemura et al., 2014).

Unlike matrix polysaccharides, cellulose is made at the plasma membrane by cellulose synthase complexes (CSCs) and CESAs activities facilitate the polymerization of the glucan chain across the plasma membrane (Purushotham et al., 2020). Prior to cellulose formation, CESAs must traverse the endomembrane system, where they are synthesized, folded, modified, and processed in the ER, Golgi, and TGN (Hoffman et al., 2021). BiFC experiments showed that CESAs interact in the ER and may assemble to form multi-protein complexes (Carroll et al., 2012) and move rapidly through the ER but slowly through the Golgi apparatus. However, their time span at the plasma membrane can be less than 15 minutes (Paredez et al., 2006), implying that a tight molecular control over the release of mature CSCs from the Golgi to the plasma membrane may exist (Sampathkumar et al., 2013).

book Objectives

This project was aimed at elucidating the functional role of three β-*GLCAT* genes (*GLCAT14A*, *GLCAT14B* and *GLCAT14C*) involved in the biosynthesis of type II arabinogalactan and their roles in growth and development of Arabidopsis. Previous work showed that *GLCAT14A*, *GLCAT14B* and *GLCAT14C* genes corresponding to *At5g39990*, *At5g15050* and *At2g37585* demonstrated β-GLCAT activity when expressed in *Pichia pastoris* by adding GlcA to both β-1,6- and β-1,3-galactan (Dilokpimol and Geshi, 2014). Also, the paired ends of GlcA have the ability to bind and release calcium in a pH dependent manner (Lamport and Varnai, 2013). Prior to the beginning of this research, the only GLCAT that was characterized extensively was ATGLCAT14A, which showed altered levels of GlcA, with a marked increase in Gal and a decrease of Ara in *atglcat14a* mutants (Knoch et al., 2013). This work was extended further to understand the biological functions of these *GLCAT14* genes (*GLCAT14A*, *GLCAT14B* and *GLCAT14C*) during vegetative growth and reproductive development of Arabidopsis using genetic mutant analysis. Specifically, my objectives in this project are to: 1) Conduct system identification and characterization of GT14 family members using bioinformatics, 2) Generate homozygous single (*glcat14a-1*, *glcat14a-2*, *glcat14b-1*, *glcat14b-2*, *glcat14c-1*, *glcat14c-2*) and double (*glcat14a/glcat14b*, *glcat14a/glcat14c* and *glcat14b/glcat14c*) and triple (*glcat14a/glcat14b/glcat14c*) *glcat14* knock-out mutants in Arabidopsis and validate the mutants using PCR and qRT-PCR; 3.) Isolate AGPs from Arabidopsis tissues (leaf, stem and siliques) from *glcat14* mutants and WT and analyze them for sugar composition and calcium content; and 4) Phenotypically characterize single and higher order *glcat14* mutants to elucidate their biological roles in growth and development of Arabidopsis.

This book has been organized into six chapters. **Chapter 1** contains a literature review covering plant cell wall structure, assembly, function, and biosynthesis. A detailed review of past research on AGP structure and biosynthesis was discussed. **Chapter 2** deals with the systems identification and characterization of GT14 family members in 14 plant genomes to understand their evolutionary relationships, sequence characteristics and gene expression profiling of GT14 family members. This chapter was published in *Scientific Reports* and addressed **objective 1**. **Chapter 3** examined the role of GLCAT14A and GLCAT14C in mucilage formation, assembly, and organization to better understand the dynamics involved in cell wall assembly. This work was published in *BMC Plant Biology* and addressed **objectives 2** and **4**. **Chapter 4** extended a previous study on GLCATs (Zhang et al., 2020). I investigated the biochemical and physiological roles of the *GLCAT14A*, *GLCAT14B* and *GLCAT14C* genes by conducting a comparative study involving the WT and single and higher order *glcat14* mutants. This work was published in *Plants* and addressed **objectives 2** and **3**. **Chapter 5** involves elucidating the role of AG glucuronidation in the sexual reproduction of Arabidopsis by conducting an in-depth assessment of the contributory role of *GLCAT14A*, *GLCAT14B* and *GLCAT14C* genes in Arabidopsis sexual fertilization and reproductive development. This chapter is currently in review by a scientific journal and addresses **objectives 2, 3** and **4. Chapter 6** provides a summary and overall conclusion with respect to the book research chapters and outlines potential future research directions based on this book.

Chapter 2: Systems Identification and Characterization of β-Glucuronosyltransferase Genes Involved in Arabinogalactan-Protein Biosynthesis in Plant Genomes

The work in this chapter was published and corresponds to the following journal citation:

Ajayi, O.O., Showalter, A.M. (2020) Systems identification and characterization of β-glucuronosyltransferase genes involved in arabinogalactan-protein biosynthesis in plant genomes. *Sci Rep* **10,** 20562.

Introduction

Plants have continually evolved adaptive features after moving to land about 400 million years ago in order to be physiologically and functionally suited for their terrestrial habitats (*Amborella* Genome Project, 2013; Zhu, 2016). The evolution of gene families can facilitate gene family expansion or contraction and is an important mechanism that drives natural variation for adaptation or speciation in green plants (Lynch and Conery, 2000). Important evolutionary mechanisms such as gene duplication and whole genome duplication events (polyploidy) and segmental gene deletion has altered gene family sizes across lineages and advanced phenotypic diversity across the plant kingdom (Tautz and Domazet-Loso, 2011). Following gene duplication and loss, adaptation and speciation appear to proceed through a combination of both structural and *cis*-regulatory changes in one or more paralogous genes (Coyne and Hoekstra, 2007). Recent advances in sequencing technology have enabled researchers to make significant progress in understanding gene evolution. This provides insights into the underlying connections between the expansion of gene families and evolution of new gene functions

(Harris and Hofmann, 2015).

Arabinogalactan-proteins (AGPs) are plant cell wall glycoproteins with

structurally complex, large-branched polysaccharides attached to hydroxyproline

(Hyp) residues. They are ubiquitous in bryophytes and higher plants (Seifert and Roberts,

2007; Ellis et al., 2010) and are implicated in virtually all aspects of plant growth and

development (Van Hengel and Roberts, 2002; Cheung et al., 1995; Knoch et al., 2014).

AGPs contain approximately 10% protein and 90% sugar, which forms the interactive

molecular surface of AGPs which are essential to their biological functions (Kieliszewski

and Lamport, 1994; Borner et al., 2003; Xu et al., 2008). Type II arabinogalactan (AG)

polysaccharides of AGPs possesses structural characteristics mainly a β-(1$\rightarrow$3)-galactan

backbone with β-(1$\rightarrow$6)-linked galactan side chains which are further decorated with

galactopyranosyl (Gal*p*) and arabinofuranosyl (Ara*f*) residues as well as with other less

abundant residues including rhamnosyl (Rha), fucosyl (Fuc), and glucuronosyl (GlcA;

with or without 4-*O*-methylation) residues (Tan et al., 2010; Tryfona et al., 2010). The

formation and assembly of AG sugar moieties are controlled by a glycosylation process

catalyzed by glycosyltransferases (GTs). GTs regulate the sequence and length of the AG

chains and act in a regio- and stereo-specific manner (Lairson et al., 2008). It is

speculated that at least ten functionally distinct GTs are required for the biosynthesis of

type II AG (Knoch et al., 2014) and many of these GTs await functional discovery and

characterization.

β-glucuronosyltransferases (GLCATs) are involved in the transfer of glucuronic acid (GlcA) to AGPs. Out of the eleven putative β-glucuronosyltransferases (GLCATs) previously discovered in *Arabidopsis thaliana* and assigned to GT14 in the Carbohydrate Active Enzyme (CAZy; database, only three GLCATs, namely, AT5G39990 (ATGLCAT14A), AT5G15050 (ATGLCAT14B) and AT2G37585 (ATGLCAT14C) have demonstrated GLCAT catalytic activity based on an *in-vitro* assay (Knoch et al., 2014) and contain the conserved Branch domain PF02485 (referred to as GLCAT domain henceforth) present in GT14 gene family. The number and genetic features of putative GLCATs in other plant species remain to be verified despite their fully sequenced genomes. Previous work using nuclear magnetic resonance and molecular dynamics simulations demonstrated that the terminal carboxyl groups of GlcA residues can act as potential intramolecular Ca^{2+}-binding sites (Lamport and Varnai, 2013). Apart from the fact that GlcA residues are the only acidic sugars in the AG chain, their importance in the binding and release of extracellular calcium have been demonstrated (Lamport and Varnai, 2013). Intuitively, GLCATs may play an essential role in global Ca^{2+} signaling processes in plants and could be a potential target for improving AGP-based products and increasing biomass for biofuel production from bioenergy crops.

The advent of high-throughput technologies in the post-genomic era has generated enormous amounts of data that can yield important biological information for researchers through data mining of publicly available databases (Showalter et al., 2010). A bioinformatics approach can be an effective tool in understanding gene family expansion, identifying paralogs and orthologs using sequence features, identifying sites under selection pressure and elucidating gene expression dynamics among gene family members. Examination of such data can yield valuable insights that can facilitate and guide further research in the field. Such is the case here, where the availability of whole genome sequences along with microarray, proteome and transcriptome data can enable large-scale investigations into identifying and characterizing gene family members in multiple plant genomes using bioinformatics approaches.

The present study is focused on mining plant genomes for evolutionary footprints of functional importance among putative GLCATs. We carried out a comprehensive analysis aimed at identifying and characterizing the family of enzymes (GLCAT) that transfer GlcA to AG glycan belonging to GT14 family using *Arabidopsis thaliana* GLCAT domain of functionally characterized GLCATs as queries. Our specific goal is to utilize phylogenetic analysis, physical mapping of genes, motif identification, synteny and gene expression analyses to gain insight into the evolution, sequence diversification, functional similarity/divergence, and tissue-specific transcriptional profiling among putative *GLCAT* gene family members in 14 plant genomes.

Materials and Methods

Sequence Retrieval and Identification of GLCAT Gene Family in Plant Lineages

Database searching was conducted with the Phytozome database version v12.1.6 (https://phytozome.jgi.doe.gov/pz/portal.html) for 14 plant genomes including *Amborella trichopoda* using the *Arabidopsis thaliana GLCAT* gene family as queries in a BLAST search. The database search was conducted for the following investigated genomes belonging to the following species: *Selaginella moellendorffii, Physcomitrella patens, Arabidopsis thaliana, Arabidopsis lyrata, Brachipodium distachyon, Citrus sinensis, Glycine max, Gossypium raimondii, Oryza sativa, Populus trichocarpa, Solanum lycopersicum, Sorghum bicolor, Vitis vinifera,* and *Zea mays.* For the identification and analysis of *GLCAT* genes in the plant genomes, genomic sequences, coding sequences (cds) and peptide sequences were downloaded from the Phytozome version v12.1.6 database (Goodstein et al., 2012) after a BLASTp 2.2.28 + search. BLASTp was performed by using *Arabidopsis GLCAT* gene members as queries to retrieve *GLCAT* gene family members of the remaining 13 investigated genomes with an *E*-value of 10^{-5}. Unique *GLCAT* genes were filtered by excluding partial and redundant sequences. All identified putative GLCAT proteins were further confirmed for the presence of family specific conserved domains using NCBI's Conserved Domain Database (CDD); (https://www.ncbi.nlm.nih.gov/Structure/cdd/wrpsb.cgi), Branch domain (PF02485) and EMBL InterProScan (https://www.ebi.ac.uk/interpro/). After confirming the presence of the conserved Branch domains (PF02485) for all putative GLCAT proteins, we further obtained gene IDs, functional annotations, chromosome locations, chromosome numbers, genomic coordinates and peptide sizes from the Phytozome database. Protein molecular weights and pI-values for the identified proteins were calculated using the ExPASy online tool (Gasteiger et al., 2005).

Phylogeny, Gene Structure Analysis, Physical Mapping and Synteny Analysis

The GLCAT protein sequences of the investigated genomes were aligned in ClustalW and illustrated using Jalview (Waterhouse et al., 2009). PhyML and

Bayesian inference (BI) were used to construct respective phylogenetic trees. PhyML was constructed using maximum likelihood with a bootstrap value of 1000 iterations, and all positions containing gaps and missing data were excluded in order to achieve phylogenetic trees. For the BI, phylogenetic reconstruction was carried out using Bayesian Markov Chain Monte Carlo (MCMC) as implemented in BEAST software v1.5.4 (Drummond and Rambaut, 2007). The analysis was carried out with the following parameters: relaxed molecular clock with an uncorrected log-normal distribution model for rate of variation, the HKY substitution model, four gamma categories and a Yule model of speciation. Three independent runs were carried out, each with 1 million MCMC generations and sampled every 1000th generation. Finally, the trees were visualized and managed in iTOL (Letunic and Bork, 2016).

A physical map of *GLCAT* gene members was constructed using the chromosome numbers and genomic coordinates in Mapchart 2.30 (Voorrips, 2002). Gene Structure Display Server (Hu et al., 2015) was used to determine the number of introns and exons. Also, syntenic regions were identified between *A thaliana* and its orthologs in the investigated genomes using Circoletto (Darzentas, 2010). The color represented the extent of similarity, blue for the lowest similarity, followed by green, orange and red, showing the increasing extent of similarity with increasing bit score. Specifically, blue for the first (i.e. worst) 25% of the maximum bitscore, green for the next 25%, orange for the third, and red for the top (i.e., best) bitscores of between 75 and 100% of the maximum bitscore (Darzentas, 2010).

Motif Identification in GLCAT Sequences in Plant Genomes

For motif identification, conserved motifs among GLCATs were identified using the MEME program (Bailey et al., 2009) with the following parameters: number of repetitions = zero or one, maximum number of motifs = 6, and optimum motif width constrained between 6 and 50 residues.

Positively Selected Sites in GLCATs and their Putative Biological Significance

For the identification of positively selected sites, we considered ATGLCAT14A (AT5G39990) orthologs since this protein is the only GLCAT that has been extensively characterized (Knoch et al., 2014) and would be a more reliable estimates of selection pressure. We performed a strict statistical analysis using the CodeML program in the EasycodeML software (Gao et al., 2019) using branch model, site model, and branch-site model (Guindon et al., 2004) in a run based on the non-synonymous (dN) and synonymous (dS) nucleotide substitution rate ratio (dN/dS) or ω. If $\omega > 1$, then there was a positive selection on some branches or sites; $\omega < 1$ suggests a purifying selection (selective constraints); and $\omega = 1$ indicates neutral evolution. The parameter estimates (ω) and likelihood scores (Wong et al., 2004) were calculated for three pairs of models. These were M0 (one-ratio) versus M3 (discrete), M1a (nearly-neutral) versus M2a (positive-selection), and M7 (beta) versus M8 (beta&ω) (Yang et al., 2005). The LRT (Nielson and Yang, 1998) was used to compare the fit to the data of two nested models, which measured the statistical significance of each pair of nested models based on the estimated *p* values (Whelan, 1999). A significantly higher likelihood of the alternative model compared to the null model suggests positive selection ($\omega > 1$). Similarly, we also implemented the branch model to select the statistically significant "foreground branch" presumed to be under positive selection while all other branches in the tree were "background" branches (for example, making branch I foreground while branch II and III are background). The background branches share the same distribution of ω values among sites, whereas different values can apply to the foreground branch. Then, the branch-site model was applied, which further estimated the different dN/dS values among the significant branches detected by the branch model and among sites (Zhang et al., 2005). Finally, a Bayes empirical Bayes (BEB) approach was then used to calculate the posterior probabilities that a site comes from the site class with $\omega > 1$, which when implemented in EasycodeML software, were used to identify sites under positive selection or purifying selection in the foreground group with significant LRTs (Nielson and Yang, 1998). Each branch group was labeled as a foreground group while other branch groups were considered background.

Digital Expression Analysis of Plant Genomes

Publicly available mRNA-seq and Affymetrix microarray data were used to determine the expression of the *GLCAT* gene family members in *Oryza sativa*, *Arabidopsis thaliana* and *Glycine max*. Complete *GLCAT* gene expression data were only available for these three species. We used the GENEVESTIGATOR software (https://genevestigator.com/gv/) because the expression dynamics for *GLCAT* genes are useful for species comparisons under different conditions. Specifically, mRNA-seq data were used to evaluate the expression of *GLCAT* gene family members across developmental stages and anatomical parts, while the Affymetrix microarray data were used to investigate the expression of the *GLCAT* gene family members during germination and abiotic conditions. Both mRNA-seq and Affymetrix data were derived from GENEVESTIGATOR (https://genevestigator.com/gv/) and species expression values for the *GLCAT* gene members were displayed as heatmaps by GENEVESTIGATOR.

Results

Sequence Retrieval, Identification and Multiple Sequence Alignments of Putative GLCAT Gene Family Members in Plant Lineages

A total of 161 putative *GLCAT* genes were found distributed across 14 plant genomes. Eleven genes were identified as members of the *GLCAT* genes in *Arabidopsis thaliana*, *Arabidopsis lyrata*, *Brachipodium distachyon*, *Oryza sativa*, *Amborella trichopoda* and *Sorghum bicolor* (sorghum). Thirteen *GLCAT* genes were identified in *Citrus sinensis* (orange) and *Populus trichocarpa* (poplar), 22 in *Glycine max* (soybean), 15 in *Solanum lycopersicum* (tomato) and *Gossypium raimondi* (cotton), 6 in *Physcomitrella patens*, 2 in *Selaginella moellendorffii*, 9 in *Vitis vinifera* (grape) and 10 in *Zea mays* (corn) (**Appendix A: Supplementary Table 2-1**). Selaginella xylosyltransferase (XYLT) genes appear to be misclassified as it contains sequences that are closer to putative GLCATs than XYLTs.

In a blastp analyses, Selaginella sequences had 60.2% sequence identity and 98.8% sequence coverage with ATGLCAT14A compared to 28% sequence identity and 69% coverage with mammalian XYLTs. Also, only the GLCAT domain is present in Selaginella and lacks the XYLT domain present in mammalian GT14 sequences

(Gotting et al., 2004). Each of these genes were confirmed for the presence of the GLCAT domain while excluding sequences having a DUF 266 domain (Ye et al., 2011). Multiple alignment of the putative GLCAT protein sequences of representative members of the investigated species identified a widely conserved GLCAT domain which is the catalytic region with less conserved N-terminal and C-terminal ends following protein sequence alignments (**Figure 2-1**). The DXD motif is a metal ion binding motif present in many GTs and this motif is essential to their catalytic functions (Gotting et al., 2004). There are some representative sequences that appear to have a highly conserved DWD motif with the first D residue less conserved than the second D residue of the motif. In addition, the second D residue of the X/DWD/X motif is substituted with either asparagine (N) or threonine (T) for some dicot species and serine (S) for some monocot grass species (**Figure 2-2B and 2-2C**). Surprisingly, we discovered a highly conserved, uncharacterized tryptophan (W) residue in the X/DWD/X motif shared among plant and animal species (**Appendix B: Supplementary Figure 2-1**).

Figure 2-1

Protein Alignment of Representative GLCATs from Various Plant Species

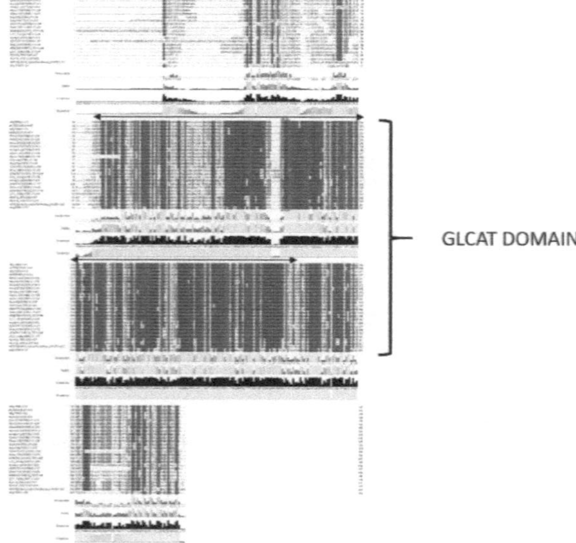

Note. Protein sequences representing the diversity of putative GLCATs in the investigated genomes were aligned using ClustalW and illustrated in Jalview. Conserved GLCAT domains are indicated with black arrow lines. The X/DWD/X motif is indicated by the red box region.

Figure 2-2

Phylogenetic Analyses of Putative GLCATs in Plant Genomes Using Maximum Likelihood Method

A

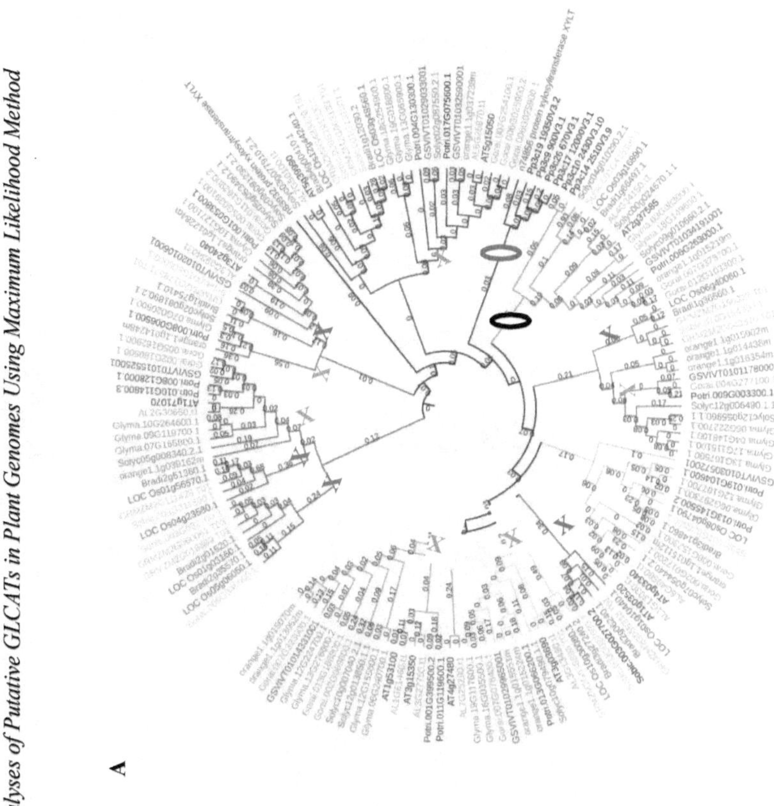

Figure 2-2: continued

B

Figure 2-2: continued

C

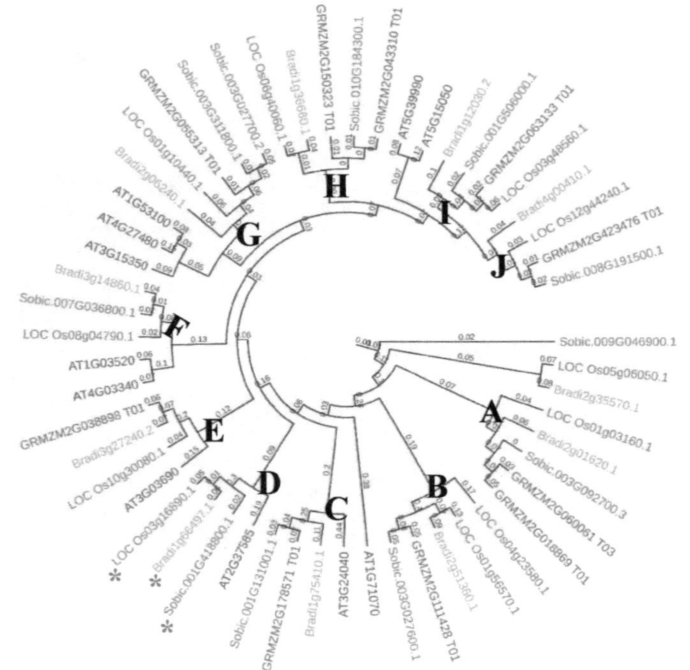

Figure 2-2: continued

D

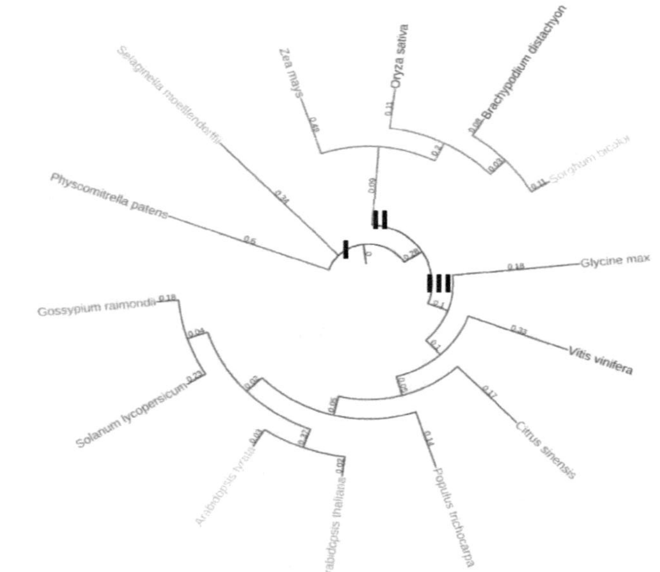

Note. Figure showed the phylogenetic analyses in 14 plant species investigated (**A**), dicots (**B**), monocot grasses (**C**) and AtGLCAT14A orthologs in 14 plant species (**D**). In (**A**), the phylogenetic tree was created with the full-length GLCAT protein sequences from 14 plant species based on the maximum likelihood method. Ten phylogenetic classes (classes **A-J**) were identified.

The blue X and red X represent dicot and monocot grass specific clades respectively, corresponding to the different species. The red oval indicates a phylogenetic clade specific to lower land plants while the black oval indicates a clade shared by lower (bryophytes) and higher land plants (monocot grass and dicots). In (**B**), eight phylogenetic clades (classes **1a**, **1b** and **2a–f**) were identified. The red and blue asterisks indicated amino acid changes in the second D residue from D → N and D → T in the DWD consensus motif found in plants respectively. In (**c**), ten phylogenetic clades (clade **A–J**) were identified and clades were assigned if it contained minimum of 3 out of the 4 monocot grasses investigated. The red asterisks indicated amino acid changes D → S in the X/DWD/X motif respectively. Arabidopsis GLCAT sequences were included for efficient classification. Genes with no asterisk possess no substitution. In (**D**), the phylogenetic tree was created with the full-length protein sequences of AtGLCAT14A (AT5G39990) and its orthologs in the 14 plant species by the Bayesian Inference using the Bayesian Markov Chain Monte Carlo method as implemented in BEAST software v1.5.4. Support values/posterior probabilities and branch lengths are indicated while I, II and III corresponds to labels for foreground and background in branch and branch-site analyses.

Phylogenetic Analyses of GLCAT Genes in the Investigated Plant Genomes

The maximum likelihood (ML) method was used to examine the phylogenetic relationships among all 14 plant species (**Figure 2-2A**), including eight dicot species (**Figure 2-2B**) and four monocot grass species (**Figure 2-2C**), which were also examined separately. Similarly, Bayesian Inference (BI) was used to assess the phylogenetic relationships among AtGLCAT14A orthologs and results were compared with those obtained using ML (**Figure 2-2D**). Results obtained from the phylogenetic tree for all species identified a phylogenetic clade (clade E) that is representative of lower and higher land plants (**Figure 2-2A**). Results obtained from the phylogenetic tree for dicot species identified two major clades, namely clades 1 and 2. Clade 1 was subdivided into groups A and B and clade 2 was subdivided into groups A–F (**Figure 2-2B**). Two Arabidopsis thaliana genes, *At3g24040* (group 1A) and *At1g71070* (group 1B) belong to clade 1 while the remaining nine Arabidopsis thaliana genes are present in clade 2. All eight dicot species have orthologs represented in each group (1A-B, 2A-F) except for group 2D which lacks *GLCAT* genes from *Arabidopsis thaliana* and *Arabidopsis lyrata*. No putative GLCAT gene members from a species were clustered into a single group except in *Amborella trichopoda*, whose putative GLCATs occupy distinct positions in the phylogeny analysis (**Appendix B: Supplementary Figure 2-6**). Clades A, B, G, H, I and J contain all species representative of monocot grass species while clades C, D and E clustered with putative *Arabidopsis thaliana GLCAT* genes (**Figure 2-2C**). In class D, the *LOC_Os03g16890.1, Bradi1g66497.1* and *Sobic. 001g418800.1* genes exhibited D to S residue changes in the second D residue of the DWD motif; this residue change is absent in its closest Arabidopsis homolog *AtGLCAT14C* (*At2g37585*) (**Figure 2-2C**).

Results of the BI analysis of *AtGLCAT14A* and its orthologs showed a similar clustering pattern that separates monocot grasses from dicots (**Figure 2-2D**).

Physical Mapping, Gene Structure Analysis and Synteny Analysis

Physical Mapping of Putative GLCAT Genes in the 14 Plant Genomes. The physical maps of putative *GLCAT* genes were distributed on all 5 chromosomes in *Arabidopsis thaliana* and seven of the eight chromosomes in *Arabidopsis lyrata.* The largest numbers were observed on chromosome 1 and 3 for *Arabidopsis thaliana* and chromosome 3 for *Arabidopsis lyrata. Solanum lycopersicum* and *Gossypium raimondi* have the same number of *GLCAT* genes distributed across 8 chromosomes in *Solanum lycopersicum* and 9 chromosomes in *Gossypium raimondi.* The largest number of genes were found on chromosomes 12 and 7 for *Solanum lycopersicum* and *Gossypium raimondi* respectively. Similarly, *Vitis vinifera* and *Glycine max* *GLCAT* genes were distributed over 7 and 12 chromosomes respectively, with the largest number of genes found on chromosomes 1 and 8 for *Vitis vinifera* and chromosomes 6, 12 and 13 for *Glycine max.* Also, putative *GLCAT* genes were distributed over 2 and 6 chromosomes in *Selaginella moellendorffii* and *Physcomitrella patens* respectively **(Appendix A: Supplementary Table 2-1 and Appendix B: Supplementary Figure 2-2).**

Gene Structure and Synteny Analyses of Putative GLCAT Genes in the 14 Plant Genomes

Gene structure diversification was studied to gain insights into the evolution of putative *GLCAT* genes in the investigated genomes. Generally, all of the *GLCAT* genes including those of the *Amborella trichopoda* have 4 exons and 3 introns with variable lengths in the 5′ and 3′ UTR regions, except for some representative species of *Glycine max*, *Gossypium raimondi*, *Physcomitrella patens* and *Citrus sinensis* (**Figure 2-3 and Appendix B: Supplementary Figure 3**). Comparative synteny relationship maps displayed a high degree of sequence similarity between *Arabidopsis thaliana GLCAT* gene family members and their respective homologs in other species (**Appendix B: Supplementary Figure 4**). Surprisingly, *At2g37585* displayed weak sequence similarity with its closest homolog in *Arabidopsis lyrata* (*Al4g34150*) (**Figure 2-4**).

Figure 2-3

Gene Structure Display of Representative GLCATs from Various Plant Species

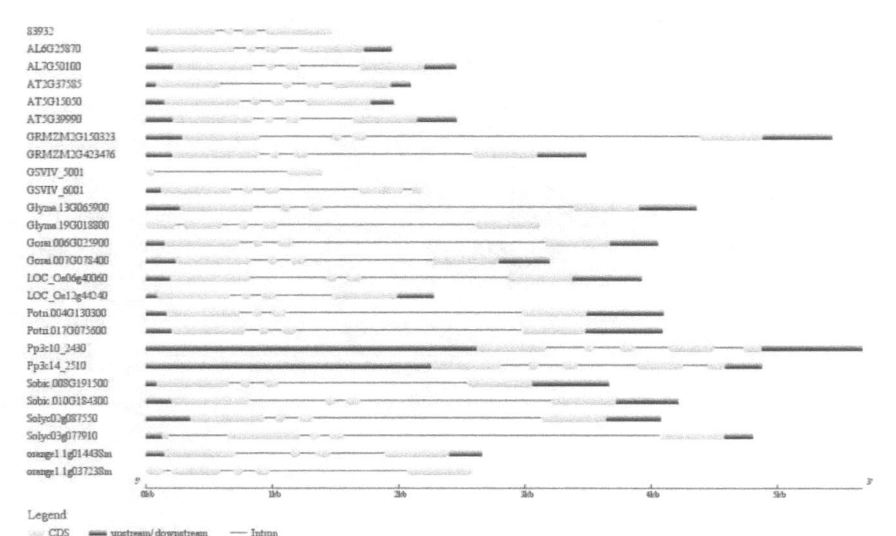

Note. Twenty-six proteins representing the diversity of putative *GLCATs* in the plant kingdom were used for the gene structure analysis as illustrated by using the GSDS 2.0 server. '83,932 = putative *GLCAT* sequence in *Sellaginella moellendorffii*; AL = *Arabidopsis lyrata*; AT = *Arabidopsis thaliana*; GRMZM = *Zea mays*; GSVIV = *Vitis vinifera*; Glyma = *Glycine max*; Gorai = *Gossypium raimondii*; LOC = *Oryza sativa*; Potri = *Populus trichocarpa*; Pp = *Physcomitrella patens*; Sobic = *Sorghum bicolor*; Solyc = *Solanum lycopersicum*; orange = *Citrus sinensis*.

Figure 2-4

Comparative Analyses of ATGLCAT14C (AT2G37585) and AL4G34150 Genes

A

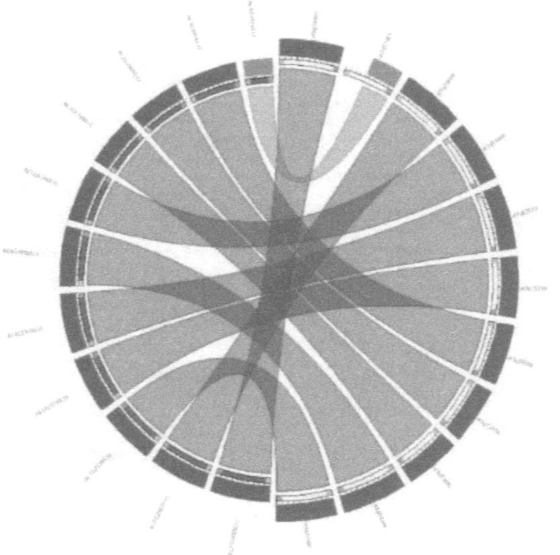

Figure 2-4: continued

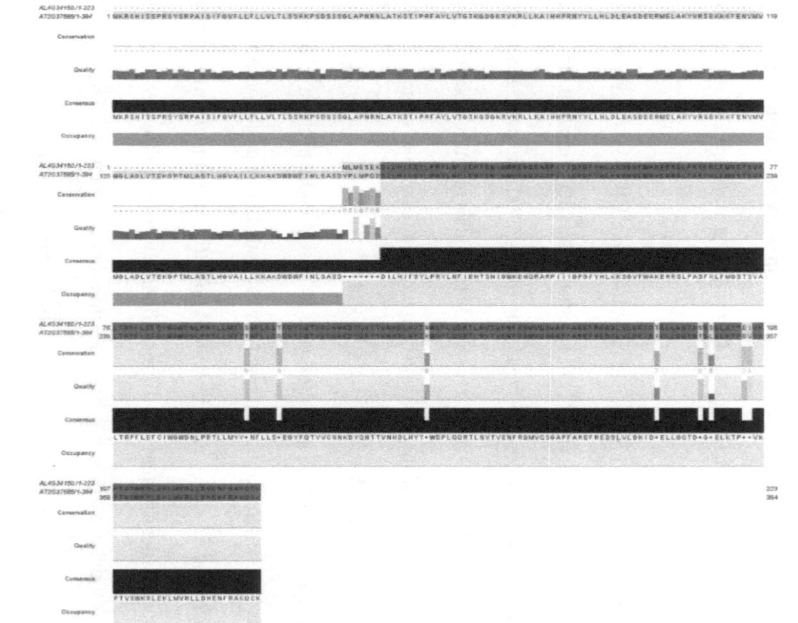

Note. (A), represents the synteny analysis of *Arabidopsis thaliana and Arabidopsis lyrata GLCAT* genes. Weak sequence similarity (green ribbon) between *ATGLCAT14C* and its ortholog *AL4G34150* in *A. lyrata.* (B) Pairwise comparison between *ATGLCAT14C* and its ortholog *AL4G34150* displaying conserved regions (highlighted in dark blue and yellow blocks) as illustrated by Jalview. The first 167 amino acid residues in *ATGLCAT14C* are conspicuously absent in *AL4G34150.*

*Distribution of Amino Acid Motifs/Blocks in Putative GLCAT Gene Family
in Plant Genomes*

A total of 161 protein sequences from 14 plant genomes was used to generate a
MEME-driven search for conserved amino acid motifs named "blocks". The MEME
analysis generated 50 sequence blocks with various lengths ranging from six (blocks 24
and 40) to 41amino acids (blocks 2, 4 and 37). Sequence blocks 1–6, 8 and 9 are found in
more than 95% of the sequences analyzed (**Appendix B: Supplementary Figure 2-5**).

*Positively Selected Sites in GLCAT14A Orthologs and their Putative Biological
Significance*

The site-specific, branch and branch-site models were used to detect sites under
selective pressure among *AtGLCAT14A* orthologs. After removing the gaps, all the
amino acid sequences were analyzed using the CodeML program. Results showed that
neither the M0 versus M3 or M2a versus M1a models identified positively selected sites.
Only the M8 versus M7 model identified several sites with ω values significantly greater
than 1 (M8 vs. M7, $2\Delta L = 261.53, p < 0.001$).

Eighty-four amino acid sites were identified under positive selection by the M8
alternative model, including 29 amino acid sites that have posterior probability
$(PP) > 0.95$ and 55 sites with $PP > 0.99$ (**Appendix A: Supplementary Table 2-2**). The
branch model identified no significant differences between the free ratio model and the
one ratio model in all the three branches investigated (**Figure 2-2D**). According to the
likelihood ratio tests (LRT) for the branch site model, comparing model A versus model
A null for the branches I, II and III, LRT were significantly different
$(2\Delta \ln L = 53.43, p = 0.00004$ for branch I, $2\Delta \ln L = 17.92711, p = 0.00002$ for branch II;

$2\Delta\ln L = 7.614822$, $p = 0.006$ for branch III). In the branch-site model, branch sites had PP < 0.95 based on the Bayes empirical Bayes computation while using each of the branches as foreground and the remaining as backgrounds (**Appendix A: Supplementary Table 2-2**).

Digital Expression Analysis of a Putative GLCAT Gene Family

The expression pattern among putative *GLCAT* gene family members in *Arabidopsis thaliana* showed differential gene expression across 32 anatomical parts with the highest expression observed in the radical elongation zone (**Figure 2-5A**). Across developmental stages, all Arabidopsis *GLCAT* gene family members had low to moderate expression except in the senescence stage. Interestingly, two genes *At4g27480* and *At1g53100* have very high expression levels in the senescence stage compared to other stages of development (**Figure 2-5B**). For the expression of *GLCAT* genes during germination in Arabidopsis, microarray data showed that *AtGLCAT14A (At5g39990)* was significantly upregulated and highly expressed (fold change > 2.5) at 1, 6, 12, 24 and 48 h of germination (**Figure 2-6A**). Similarly, significant upregulation (fold change > 2.5) was observed for *AtGLCAT14B (At5g15050), At3g15350, At3g24040* and *At1g71070* at 6, 12, 24 and 48hrs of germination (**Figure 2-6A**). Expression profile of *GLCAT* gene family members in *Oryza sativa* identified *sativa* identified *LOC_Os12g44240* as expressed in all 20 anatomical parts with differential expression patterns observed among rice *GLCAT* gene members (**Figure 2-5C**). Low to medium expression was observed among the genes; however, two genes, *LOC_Os3g48560* and *LOC_Os1g56570* were highly expressed in the maturation zone of rice root seedlings (**Figure 2-5C**). Differential gene expression was observed across all developmental stages (**Figure 2-5D**). For *GLCAT* gene family members in soybean across 27 anatomical parts, medium expression was observed for most soybean *GLCAT* genes except for *Glyma.03g083000* whose expression was not detected (**Figure 2-5E**). Higher expression during the seedling stage was observed in root hairs for *Glyma.12G155800*, while *Glyma.06G240700* was expressed predominantly in the elongation zone of seedlings (**Figure 2-5E**). Across developmental stages, low to medium expression was observed across *GLCAT* gene family members except for *Glyma.03G083000* which had no detectable expression (**Figure 2-5F**). For soybean *GLCAT* gene family, most of the genes were significantly upregulated (fold change > 2.5) especially at 12 h and 24 h of germination (**Figure 2-6B**). Notably, *Glyma.13g065900, Glyma.19g018800, Glyma.12g107700, Glyma.10g264600, Glyma.09g119700, Glyma.10g127100* were highly upregulated (fold change > 2.5) at 3 h, 6 h, 12 h and 24 h of germination (**Figure 2-6B**).

Figure 2-5

Expression Pattern of Putative GLCAT Genes Across Anatomical Reference Points and Developmental Stages in Arabidopsis, Rice and Soybean

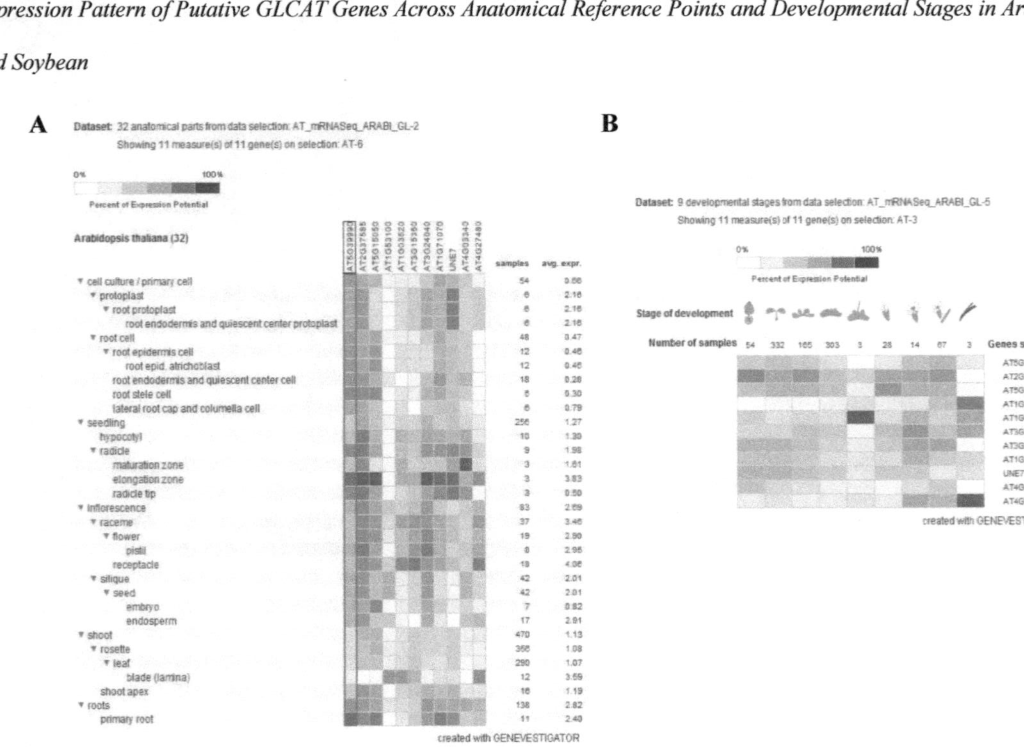

Figure 2-5: continued

C

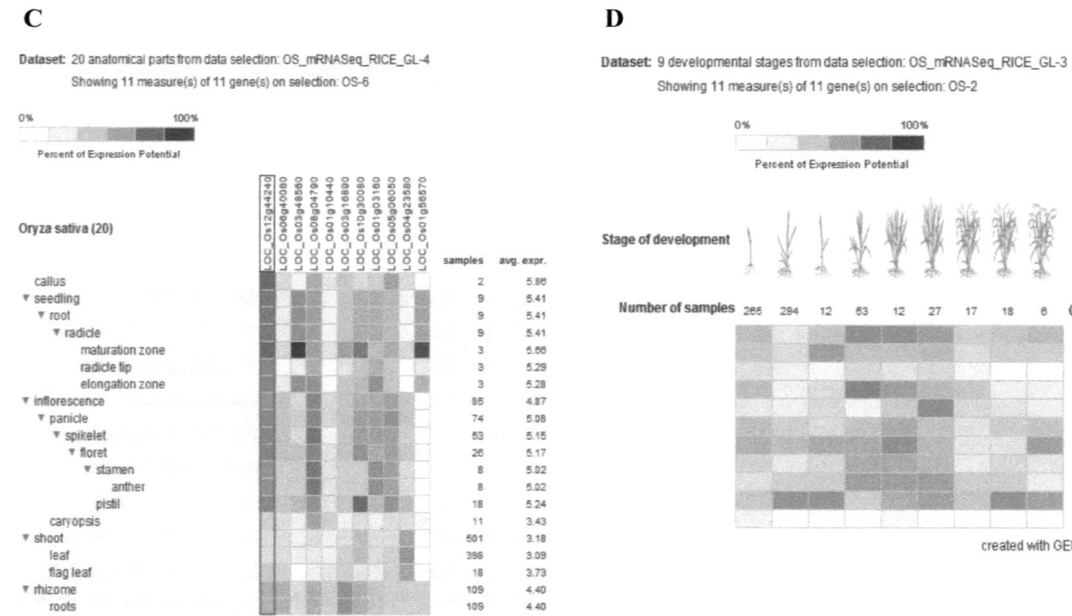

D

Figure 2-5: continued

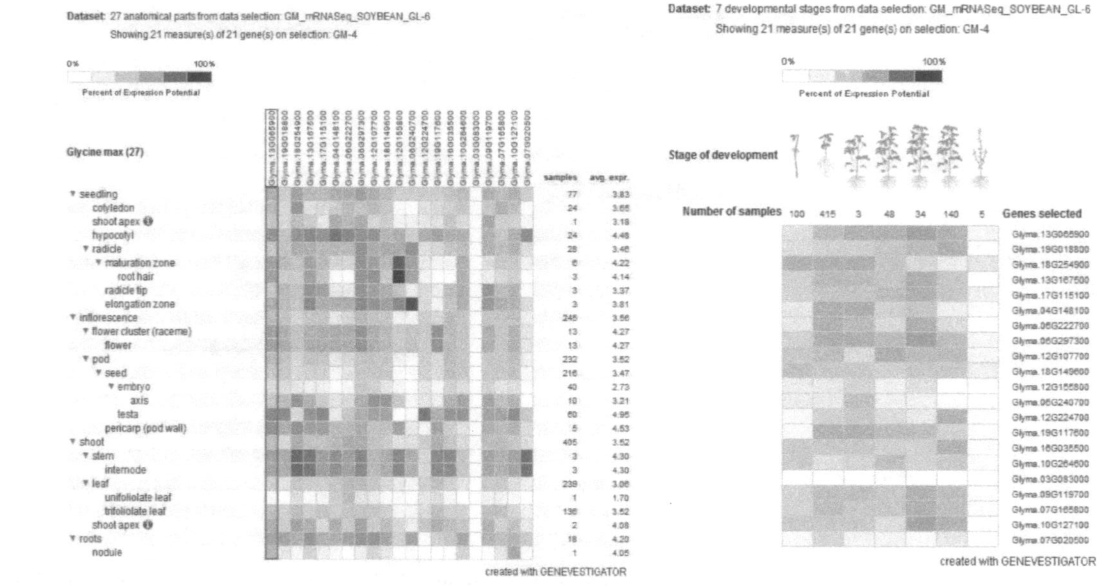

Note. Figure represents the digital expression pattern in the anatomical parts and across developmental stages of Arabidopsis (A, B), rice (C, D) and soybean (E, F). The red boxed region corresponds to the functionally characterized *AtGLCAT14A* (AT5g39990) gene (A) and its orthologs in rice (C) and soybean (E)

Figure 2-6

Digital Expression Pattern of Putative GLCATs in Arabidopsis and Soybean During Germination

A

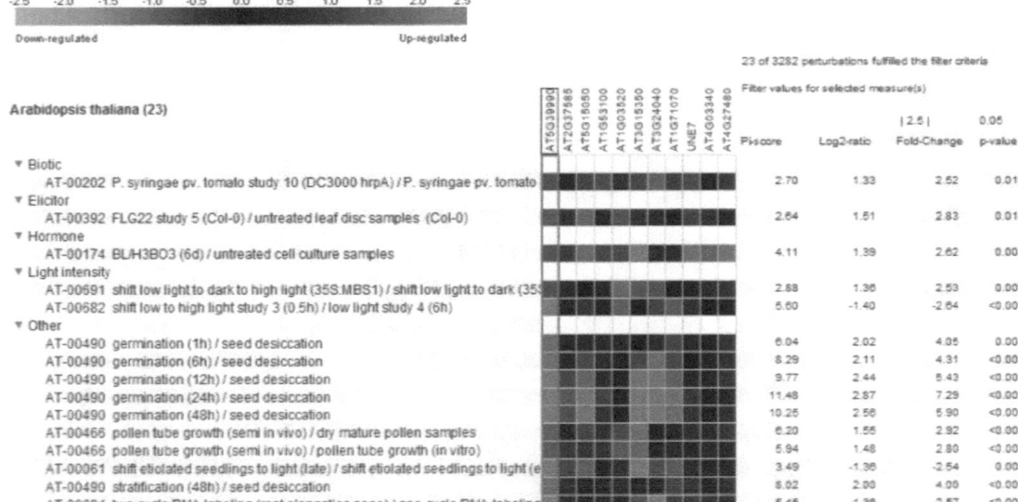

	Pi-score	Log2-ratio	|2.5| Fold-Change	0.05 p-value
▼ Biotic				
AT-00202 P. syringae pv. tomato study 10 (DC3000 hrpA) / P. syringae pv. tomato	2.70	1.33	2.52	0.010
▼ Elicitor				
AT-00392 FLG22 study 5 (Col-0) / untreated leaf disc samples (Col-0)	2.64	1.51	2.83	0.018
▼ Hormone				
AT-00174 BL/H3BO3 (6d) / untreated cell culture samples	4.11	1.39	2.62	0.001
▼ Light intensity				
AT-00691 shift low light to dark to high light (35S:MBS1) / shift low light to dark (35S	2.88	1.36	2.53	0.007
AT-00682 shift low to high light study 3 (0.5h) / low light study 4 (6h)	5.60	-1.40	-2.64	<0.001
▼ Other				
AT-00490 germination (1h) / seed desiccation	6.04	2.02	4.05	0.001
AT-00490 germination (6h) / seed desiccation	8.29	2.11	4.31	<0.001
AT-00490 germination (12h) / seed desiccation	9.77	2.44	5.43	<0.001
AT-00490 germination (24h) / seed desiccation	11.48	2.87	7.29	<0.001
AT-00490 germination (48h) / seed desiccation	10.25	2.56	5.90	<0.001
AT-00466 pollen tube growth (semi in vivo) / dry mature pollen samples	6.20	1.55	2.92	<0.001
AT-00466 pollen tube growth (semi in vivo) / pollen tube growth (in vitro)	5.94	1.48	2.80	<0.001
AT-00061 shift etiolated seedlings to light (late) / shift etiolated seedlings to light (e	3.49	-1.36	-2.54	0.003
AT-00490 stratification (48h) / seed desiccation	8.02	2.00	4.00	<0.001
AT-00694 two-cycle RNA labeling (root elongation zone) / one-cycle RNA labeling	5.45	-1.36	-2.57	<0.001

Figure 2-6: continued

B

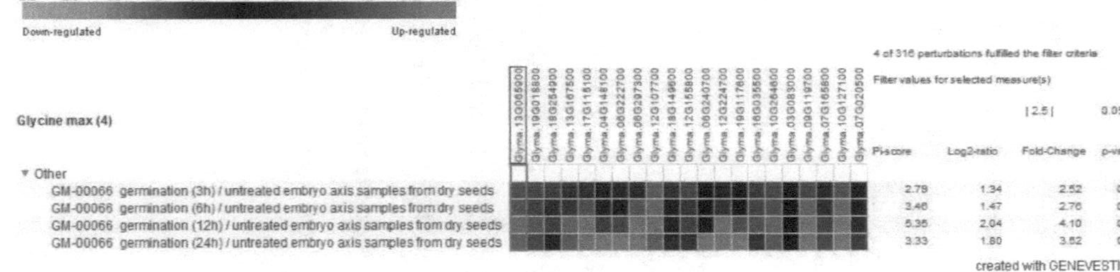

Note. (**a**) Expression pattern of *GLCAT* genes in *Arabidopsis thaliana* during germination. (**b**) Expression pattern of *GLCAT* genes in soybean during germination. The red boxed regions indicated the A*tGLCAT14A* (AT5g39990) gene and its ortholog in soybean.

Digital Expression of Putative GLCAT Genes Under Abiotic Stress in Arabidopsis thaliana and Oryza sativa

The modification of plant cell walls is one of the key processes that drives the adaptability of crop species to abiotic stressors. In heat, anoxia and hypoxia studies in *Arabidopsis thaliana*, all the *GLCAT* gene family members either showed no change in expression or were significantly down-regulated (**Figure 2-7A**). Interestingly, *AtGLCAT14A* and *At1g71070* were significantly down-regulated (fold change < 1.5) in all the stressors examined (**Figure 2-7A**). In rice, abiotic expression data was only available for cold stress in specific rice genotypes (Cold tolerance imbred line K354 and its recurrent parent C418, which possesses a cold sensitive phenotype; IR29 and LTH, both chilling-tolerant and chilling-sensitive lines, respectively). For the responses of rice *GLCAT* gene family members to cold stress, *LOC_Os12g44240* was down-regulated (fold change < 2.5) following exposure to cold stress (4 °C) for 24 h and 48 h in shoot and leaf tissues but up-regulated when re-exposed to 29 °C for 24 h only in IR29 (indica) and LTH (japonica) genotypes (**Figure 2-7B**). Surprisingly, *LOC_Os03g16890* was significantly up-regulated (fold change > 1.5) under cold stress but down-regulated during the recovery phase (at 29 °C) in both IR29 and LTH genotypes. Also, *LOC_Os04g23580* was highly up-regulated during the recovery phase (at 29 °C) in IR29 and LTH genotypes while *LOC_Os12g44240* was up-regulated only in IR29 but not in the LTH genotype (**Figure 2-7B**).

Figure 2-7

Expression Pattern of Putative GLCATs Genes in Arabidopsis and Rice in Response to Abiotic Stress

A

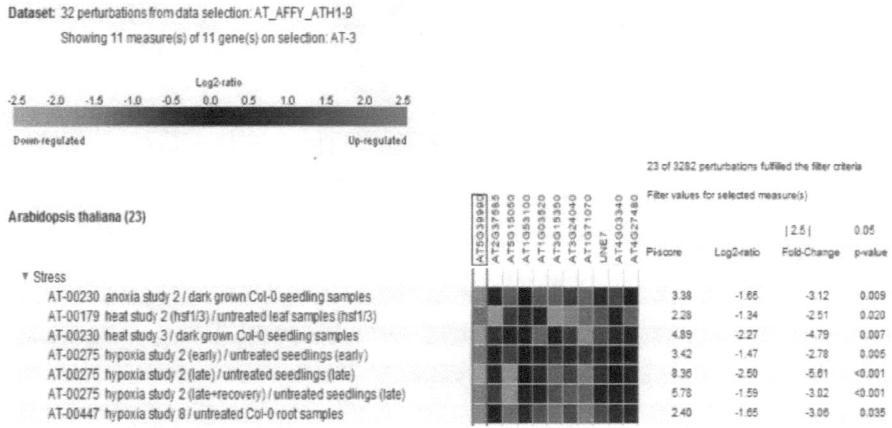
Dataset: 32 perturbations from data selection: AT_AFFY_ATH1-9
Showing 11 measure(s) of 11 gene(s) on selection: AT-3
Log2-ratio
-2.5 -2.0 -1.5 -1.0 -0.5 0.0 0.5 1.0 1.5 2.0 2.5
Down-regulated
Up-regulated
23 of 3282 perturbations fulfilled the filter criteria
Filter values for selected measure(s)
Arabidopsis thaliana (23)
AT5G39990
AT2G37585
AT5G15050
AT1G53100
AT1G03520
AT3G15350
AT3G24040
AT1G71070
UNE7
AT4G03340
AT4G27480
| 2.5 |
0.05
Pi-score
Log2-ratio
Fold-Change
p-value
Stress
AT-00230 anoxia study 2 / dark grown Col-0 seedling samples 3.38 -1.65 -3.12 0.009
AT-00179 heat study 2 (hsf1/3) / untreated leaf samples (hsf1/3) 2.28 -1.34 -2.51 0.020
AT-00230 heat study 3 / dark grown Col-0 seedling samples 4.89 -2.27 -4.79 0.007
AT-00275 hypoxia study 2 (early) / untreated seedlings (early) 3.42 -1.47 -2.78 0.005
AT-00275 hypoxia study 2 (late) / untreated seedlings (late) 8.36 -2.50 -5.61 <0.001
AT-00275 hypoxia study 2 (late+recovery) / untreated seedlings (late) 5.78 -1.59 -3.02 <0.001
AT-00447 hypoxia study 8 / untreated Col-0 root samples 2.40 -1.65 -3.06 0.035

Figure 2-7: continued

B

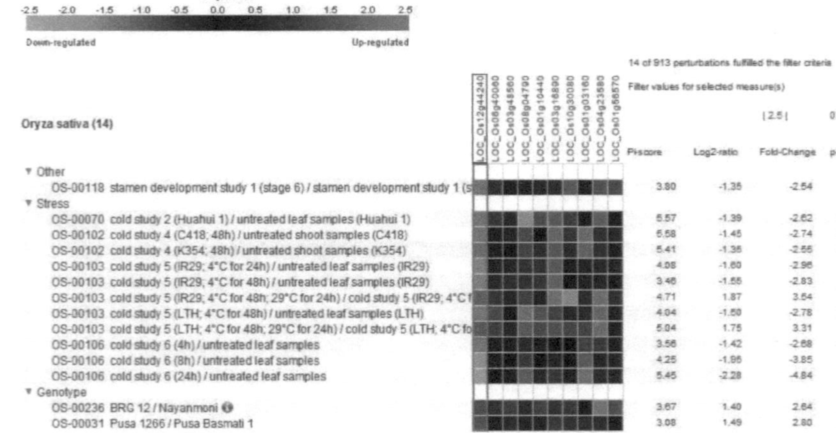

Note. (A) Expression pattern of *GLCAT* genes in *Arabidopsis thaliana* during anoxia, heat and hypoxia (B) Expression pattern of *GLCAT* genes in rice during cold stress. Note the differential gene expression in all *GLCATs* in Arabidopsis and rice in response to different stressors. The red boxed region indicates AtGlcAT14A (AT5g39990) and its ortholog in rice, both of which are relatively downregulated in response to stressors.

Discussion

AGPs are a complex family of macromolecules with heterogenous sugar chains decorating the protein backbones. AGPs are plant cell wall glycoprotein components with great potential for industrial and food applications, which builds upon their well-known emulsifying, adhesive, and water-holding properties. We considered the plant GT14 gene family as putative GLCATs for the following reasons: (1) The glucuronic acid substitution of xylan (GUX) enzymes have been demonstrated to attach α-GlcA to xylan and the loss of function of *GUX* genes resulted in the complete loss of GlcA in xylan (Lee et al., 2012); as such, we speculate that a different set of enzymes may be responsible for the AGP-specific GlcAT activity. (2) The investigated genes share similar domain characteristics with functionally characterized GLCATs (ATGLCAT14A, ATGLCAT14B and ATGLCAT14C) whose function primarily involves the transfer of β-GlcA to AGPs (Knoch et al., 2014). (3) The *glcat14aglcat14bglcat14c* genetic knock-out mutants resulted in only 40% reduction in GlcA content in AGP isolates (Zhang et al., 2020). Unlike the complete loss of GlcA in gux1/2/3 triple mutant, the 40% loss of GlcA in *glcat14aglcat14bglcat14c* triple mutant may suggest the possible involvement of other putative *GLCAT* genes in the plant GT14 family in AGP GLCAT activity. In this study, we focused on a family of enzymes called GLCATs that are involved in the transfer of GlcA to AGPs for two reasons. First, GlcA is the only acidic sugar decorating the large arabinogalactan (AG) sugar chains and is known to bind to calcium in a pH-dependent manner with functional ramifications (Lamport and Varnai, 2013). Given the important role of calcium in plant growth and developmental processes, the understanding of the evolution of this gene family will provide key insights that could be exploited for use in increasing plant biomass in bioenergy crops. Second, previous work showed that the addition of extra amounts of GlcA to commercially available gum arabic improved its emulsification properties (Geshi et al., 2013). This offers the opportunity of developing new gum arabic variants useful in the design of novel AGP-based products. To this end, we identified and characterized putative *GLCAT* gene family members in 14 plant genomes to gain insight into their evolution, sequence diversification, functional similarities/ differences, and patterns of gene expression.

Gene families include genes that share similar cellular functions and commonly arise as a result of gene or genome duplication events. Gene duplication and deletion are generally considered important evolutionary mechanisms that give rise to phenotypic diversity (Harris and Hofmann, 2015). The number of putative *GLCAT* gene family members in the investigated land plant species are comparable to an earlier report (Ye et al., 2011) and ranged from two *GLCAT* genes in *Selaginella moellendorffii* to 22 *GLCAT* genes in soybean (**Appendix A: Supplementary Table 2-1**). This may be an indication of the extent of gene family expansion of *GLCATs* across evolutionary time, conceivably driven by forces of natural selection, coupled with the demand to adapt to a life on land.

Conserved regions among gene family members across species can identify functionally important motifs. Many GTs require divalent metal ions, commonly Mn^{2+} or Mg^{2+}, for catalytic activity (Lairson et al., 2008) which bind to DXD motifs. In Arabidopsis, ATGLCAT14A showed full activity in the absence of Mn^{2+} or Mg^{2+}; however, the addition of EDTA (> 5 mM) inhibited its activity. Apparently, this suggests that other divalent metal ions may be involved in enzyme activity (Knoch et al., 2014). In this study, we identified a moderately conserved motif, X/DWD/X among putative *GLCAT* gene family members with 71 out of 161 sequences having the DWD motif, including a bryophyte and a lycophyte. Representative sequences of the mammalian *O*-β-xylosyltransferases (XT-I, XT-II), belonging to the GT14 family, possesses a highly conserved and functionally characterized DXD motif (Gotting et al., 2004) that is absent in plant GLCATs and an uncharacterized X/DWD motif common to plants and mammals (**Appendix B: Supplementary Figure 2-1**). Gotting et al. (2004) argued that the DWD motif is unique in the sense that no GT families investigated so far have an aromatic amino acid, tryptophan (W), at the central position of the consensus motif, and this observation has been reported earlier (Wiggins and Munro, 1998). Interestingly, an investigation into the role of the tryptophan residue demonstrated that alterations of the tryptophan (W) residue in the DWD motif in the human XT-1 protein to either a neutral, basic, or acidic amino acid resulted in a > 60% reduction in enzyme activity (Gotting et al., 2004). Although, this functionally important DWD motif in human XT-1 does not align with the Arabidopsis GLCAT sequences in this study (**Appendix B: Supplementary Figure 1**), we hypothesize that the highly conserved W residue in the X/DWD motif shared by all species may have been evolutionarily selected to play either a functional or structural role that promotes GLCAT catalytic activity in plants especially when this is the only X/DWD/X motif present among plant GT14 family members. Site-directed mutagenesis studies aimed at discovering the role of this conserved W and the flanking aspartate residues in the DWD motif could address this hypothesis.

Phylogenetic analysis provides a method for assessing homology and inferring relationships among genes. Ten clades (A-J) were identified in a phylogenetic analysis of all putative *GLCAT* gene family members investigated across plant species (**Figure 2-2A**). Surprisingly, we discovered that clade E family members share a common ancestor before the divergence between higher and lower land plants. It is conceivable that based on the phylogenetic analysis, *Pp3c10_2430v3.10* and *Pp3c14_2510v3* found in clade E played key roles in expansion of the *GLCAT* genes into dicots and monocot grasses (**Figure 2-2A**). Dicot specific phylogenetic analysis showed two distinct phylogenetic clades (1A and B; 2A–F), each having species representatives in each subclade (**Figure 2-2B**). With the exception of the putative *GLCAT* genes in *Amborella trichopoda* (**Appendix B: Supplementary Figure 2-6**), no single clade was found to consist of *GLCAT* genes from a single species of monocot grasses and dicots, suggesting that the putative GLCATs in *Amborella trichopoda* evolved prior to the emergence of putative GLCATs in angiosperms (Dicot and grasses).

GLCAT gene structure was highly conserved with most species having 4 exons and 3 introns in 97% of the sequences which includes lower and higher land plants. Similarly, the average number of amino acids in each species ranged from 360–450 amino acids (**Appendix A: Supplementary Table 2-1**). The above observations showed that the genetic architecture of *GLCAT* gene family members in lower and higher land plants appears to be evolutionarily conserved. In addition, we observed that the same gene architecture (4 exons, 3 introns) is present in *Amborella trichopoda* and the lower land plants examined (a bryophyte and a lycophyte). It is therefore conceivable that this gene structure originated from the shared common ancestor between the lower and higher land plants, possibly prior to the movement of plants from aquatic to terrestrial life.

Plant genomes contain a variety of structured patterns that are conserved and can be used to discover putative functions of gene family members. We identified 9 sequence blocks (blocks 1–7, 11 and 12, hence referred to as GLCAT motifs) localized in the GLCAT domain. Eight sequence blocks (blocks 1–6, 8 and 9) are present in 95% of the GLCAT sequences; however, blocks 8 and 9 are in the N-terminal and C-terminal regions flanking the GLCAT domains, respectively (**Appendix B: Supplementary Figure 2-5**). Surprisingly, block 2 which contains the highly conserved W residue of the DWD consensus motif is present in all the sequences except in *Arabidopsis lyrata* (*AL4G34150*), which after further investigation demonstrated weak sequence similarity with its closest homolog *AtGLCAT14C* (**Figure 2-4A**). Despite the fact that other *Arabidopsis thaliana GLCAT* gene family members showed high sequence similarity with *Arabidopsis lyrata* gene family members as revealed by phylogenetic analyses (**Figure 2-2B**), *AL4G34150* and *AtGLCAT14C* share two things in common: (1) they both have their exon–intron structure conserved and reside in the same phylogenetic clade (clade E) with *P. patens* (**Figure 2-2A**) and (2) they both lack some blocks found in the GLCAT domain (*AtGLCAT14C* lacks blocks 11 and 12, while *AL4G34150* lacks blocks 2, 4, 5, 7 and 11). Although, *AtGLCAT14C* (*At2g37585*) has *in-vitro* GLCAT activity (Geshi et al., 2013), it is unknown whether its closest homolog in *Arabidopsis* *lyrata* has such GLCAT activity. This raises some interesting questions. For example, since ATGLCAT14C lacks blocks 11 and 12 in the GLCAT domain, could it be that sequence blocks 11 and 12 are not critical to GLCAT activity *in vivo*? Does its closest homolog *AL4G34150*, which lacks several blocks in the GLCAT domain and lacks the highly conserved W residue in the DWD consensus motif possess GLCAT activity? We speculate that generating and expressing constructs that lack one or more of these motifs in the GLCAT domain and testing them in an *in-vitro* assay will provide answers to these questions.

Previous research reported that positively selected genes are more likely to interact with each other than genes not under positive selection (Vamathevan et al., 2008). Using the site model of BI analysis, ω values identified amino acid residues under positive selection ($p < 0.05$). The branch model did not identify a branch under positive selection, while the branch-site model identified some sites but with significance values greater than 0.05 (**Appendix A: Supplementary Table 2-2**). The inability to detect positively selected sites ($p > 0.05$) in the branch-site model is indicative of functional similarity among AtGLCAT14A orthologs in both lower and higher land plants. Similarity in the genetic architecture, the conserved GLCAT domain and the shared phylogenetic clades of *GLCAT* genes may be an evolutionary pattern that is necessary for GLCAT activity *in-vivo*. It is well documented that the *cis*-regulatory elements are seen as the most likely target for the evolution of gene regulation since the modular nature of *cis*-regulatory elements largely frees protein coding regions from deleterious pleiotropic effects (Carroll et al., 2005). We are, however, not certain that this is the case in GLCATs since we did not investigate the role of *cis*-regulatory sites and their contribution to gene function in the investigated genomes.

The availability of high throughput techniques has led to the release of large amounts of data that could be mined to understand gene expression of candidate genes of major crops. Variations in cell wall composition are linked to differential gene expression patterns in different plant tissues (Minic, 2008; McKinley et al., 2016). In the species investigated, varying degrees of gene expression of *GLCAT* gene family members were observed across developmental stages and anatomical parts. Similarly, gene expression profiles in the anatomical parts of Arabidopsis, rice and soybean showed that most of the putative *GLCAT* gene family members have moderate to high expression values during the seedling stage, primarily in the elongation and maturation zones of the radicle (**Figure 2-5A, 2-5C and 2-5E**). This might be reflective of the extensive cell wall modifications necessary for radicle growth.

We investigated the gene expression pattern of all *GLCAT* gene family members and found *AtGLCAT14A* (AT5G39990) homologs in rice (*LOC_Os12g44240.1*) and soybean (*Glyma. 13G065900*) to be consistently expressed across all anatomical parts and may reflect conserved gene regulatory processes. In contrast, varying degrees of gene expression were observed across developmental stages and may suggest some variation of GlcA abundance in their polysaccharide chains at specific stages of development (Geshi et al., 2013).

Radicle emergence from the seed is a highly regulated process that involves discrete and coordinated changes in plant cell wall extensibility and rearrangements of its components (Gómez-Maqueo and Gamboa-deBuen, 2016). The available data in GENEVESTIGATOR allows us to investigate the gene expression pattern of Arabidopsis and soybean during germination. While there are differential expression patterns across putative *GLCAT* gene family members in both species, there were some parallels in their temporal expression patterns. Notably, some genes in Arabidopsis and soybean were significantly up-regulated (fold change > 2) across all time points (**Figure 2-6A and 2-6B**). Some *GLCATs* genes have low or no expression across all or some of the time points in Arabidopsis and soybean. These two observations underscore the importance of certain *GLCAT* gene family members in germination as growing cells need to produce tailored polysaccharide-rich cell walls essential to determining overall plant growth and biomass.

As sessile organisms, plants must cope with the onslaught of abiotic stresses, such as cold, hypoxia, and extreme temperatures. As most of the *GLCAT* genes were significantly up-regulated during germination, some of the *GLCAT* gene family members were significantly down-regulated in response to abiotic stress. We observed that the expression of most, if not all of the *GLCAT* genes in Arabidopsis were either unchanged or severely down-regulated (fold change < 2) in response to anoxia, hypoxia and heat stress (**Figure 2-7A**). This might be as a result of the need to divert resources from growth to the expression of genes involved in adaptation and survival. In all the rice genotypes examined, we observed a differential response to cold stress; while most *GLCAT* genes were down-regulated, *LOC_Os03g16890* gene was consistently up-regulated in response to cold (4 °C) but down-regulated as the temperature increased to 29 °C in the LTH and IR29 rice genotypes.

Notably, *LOC_Os12g44240* (an *AtGLCAT14A* homolog) was consistently down-regulated in response to cold, but up-regulated as the temperature increased to 29 °C in the LTH and IR29 rice genotypes (**Figure 2-7B**). Taken together, there appear to be differences in *GLCAT* gene expression in response to abiotic stress, with concomitant effects on plant growth and biomass production.

Chapter 3: Two β-Glucuronosyltransferases Involved in the Biosynthesis of Type II Arabinogalactans Function

in Mucilage Polysaccharide Matrix Organization in *Arabidopsis*

The work in this chapter was published and corresponds to the following journal citation:

Ajayi, O.O., Held, M. A., and Showalter, A.M (2021) Two β-Glucuronosyltransferases Involved in the Biosynthesis of Type II Arabinogalactans Function in Mucilage Polysaccharide Matrix Organization in Arabidopsis thaliana. *BMC Plant Biology* 21:245

Introduction

Normal plant development depends critically on the interactions between different components of the plant cell wall. This dynamic structure defines the plant morphological architecture and is responsible for cell shape, cell adhesion and organ cohesion (Griffiths and North, 2017). Plant cell walls are initiated by the synthesis, secretion, modification and crosslinking of individual wall components– cellulose, hemicellulose, pectin and hydroxyproline-rich glycoproteins- and are synthesized by the coordinated action of a myriad of glycosyltransferases. Understanding the underlying mechanisms involved in the assembly of complex polysaccharide network and elucidating their biological roles is not a trivial task (Arsovski et al., 2010), and remains to date a key goal for scientists interested in the manipulation of plant cell wall structure to better understand its physiological functions and its commercial exploitation.

One model system that is gaining increasing recognition and significance for the study of cell wall polysaccharide interactions is the Arabidopsis seed coat epidermis

(SCE), also referred to as Mucilage Secretary Cells (MSC) (Griffiths and North, 2017). The SCE is an excellent model system for understanding the genetic basis of cell wall biosynthesis, secretion, assembly and modification (Haughn and Western, 2012; Voiniciuc et al., 2015) because large amounts of cell wall polysaccharides can be extracted with ease and analyzed in a short timeframe. Between 5- and 8-days post anthesis (DPA), large amounts of pectins are secreted to the apoplastic space at the junction of the outer tangential and radial primary walls, forming a donut-shaped pocket of mucilage around a cytoplasmic column (Voiniciuc et al., 2015). The epidermal cells then synthesize a volcano-shaped secondary wall (9 to 11DPA) called the columella, which protrudes through the center of the mucilage pocket and connects to the primary wall. When dry, mature seeds imbibe water, rapid mucilage expansion ruptures the tangential SCE to release the polysaccharide-rich mucilage that is organized in two distinct layers: an outer, water soluble non-adherent layer and an inner, adherent layer that remains tightly attached to the seed coat surface. Arabidopsis mucilage is composed primarily of unbranched RG-I, with small quantities of HG, cellulose, and arabinoxylan found in the inner layer (Haughn and Western, 2012; Western et al., 2000). Several attempts have been made to better understand the functional roles of the glycosyltransferases involved in cell wall biosynthesis, secretion, and delivery of the mucilage polymers through the analysis of mucilage mutants. In recent years, genetic mutants that lack functional enzymes required for mucilage biosynthesis and extrusion have been identified and characterized, but many others await functional investigation.

Work to date has identified several genes/proteins involved in mucilage biosynthesis, including a fasciclin-like arabinogalactan-protein (AGP) named SALT OVERLY SENSITIVE 5 (SOS5), GALT2 and GALT5, two galactosyltransferases responsible for initiating glycosylation of AGPs, and a receptor-like kinase called FEI2; individual as well as higher order mutants corresponding to these genes/proteins are characterized by mucilage pectin repartitioning and the marked absence of cellulosic rays, while the diffuse cellulose staining remains intact (Griffiths et al, 2016). As SOS5, which is also known as FLA4, is the only well characterized AGP reported to be involved in mucilage biosynthesis, the contribution of the SOS5 glycan moieties and potentially other AGPs to mucilage formation is far from complete and presents an enigma worth unraveling.

AGPs are a family of hydroxyproline-rich glycoproteins that are extensively glycosylated with Type II AGs that are covalently attached to hydroxyproline residues in the AGP protein backbone (Showalter, 2001; Ellis et al., 2010). An individual type II AG glycan consists of a β-1,3-galactan backbone with β-1,6-galactosyl branches that are decorated with arabinosyl residues and often with other minor sugar residues, such as glucuronic acid (GlcA), rhamnose (Rha), and Fuc (Ellis et al., 2010; Tan et al., 2012). Although their exact roles in mucilage formation are still unclear, the interaction of AGPs with wall polysaccharides, their involvement in intracellular signaling cascades, and their influence on a wide variety of biological processes are known (Ellis et al., 2010; Tan et al 2013). Notably, the complexity of the cell wall polymer network with respect to AGPs is perhaps best illustrated by the finding that AGPs form covalent linkages to both RG I and arabinoxylan (Tan et al 2013).

Three glucuronosyltransferases (GLCATs), GLCAT14A, GLCAT14B and GLCAT14C were functionally characterized and found to transfer GlcA residues to AGPs (Dilokpimol and Geshi, 2014), while two additional GLCATs (GLCAT14D and GLCAT14E) were also reported to be involved in the glucuronidation of AGPs (Lopez-Hernandez et al., 2020). Here, we present evidence that two GLCATs (GLCAT14A and GLCAT14C) belonging to family GT14 in the Carbohydrate-Active Enzymes (CAZy) classification system (http://www.cazy.org; (Cantarel et al., 2009) are critically important in mucilage matrix formation in Arabidopsis.

Materials and Methods

Plant Lines and Plant Growth Conditions

Arabidopsis thaliana accession Columbia-0 (Col-0) and two T-DNA insertion lines for *At5g39990-*(*glcat14a-1*, Salk_064313 and *glcat14a-2*, Salk_043905) and *At2g37585-* (*glcat14c-1*; Salk_005705) were obtained from the Arabidopsis Biological Resource Center (ABRC, Ohio State University). Seeds were germinated on plates with 0.5% MS media, after 4 days of stratification in the dark at 4 °C and were grown under long-day conditions (16 h of light/8 h of dark, 22 °C, 60 % humidity) in growth chambers. Seedlings were transplanted after 7 days and grown under long-day conditions (16 h of light/8 h of dark, 22 °C, 60 % humidity). The *glcat14a-1glcat14c-1* double mutant was isolated from an F2 population from a cross between the two respective single-mutant parents.

Mutant Confirmation by PCR and qRT-PCR

Mutant plants were genotyped following DNA extraction using the 2× CTAB method by utilizing gene-specific primers in conjunction with the LBb1.3 insert-specific primer (**Appendix A: Supplementary Table 3-1**) targeting specific regions as indicated in **Figure 3-1B** in a PCR. To analyze transcript levels of *GLCAT14A* and *GLCAT14C* in mutants, total RNA was extracted from siliques at the linear cotyledon stage (8DAP). RNA (1 µg) was used for first-strand cDNA synthesis along with an oligonucleotide (dT20) primer and SuperScript III reverse transcriptase (Thermo Scientific). The qPCR was performed using appropriate qPCR primers (**Appendix A: Supplementary Table 3-1**) following procedures described here (Zhang et al., 2020). Moreover, *GLCAT14A* and *GLCAT14C* expression during seed coat development was also examined using the Arabidopsis seed coat-specific expression browser (http://bar.utoronto.ca/efp_seedcoat/cgi-bin/efpWeb.cgi).

Figure 3-1

Overview of GLCAT14A and GLCAT14C Gene Expression and Mutations

A

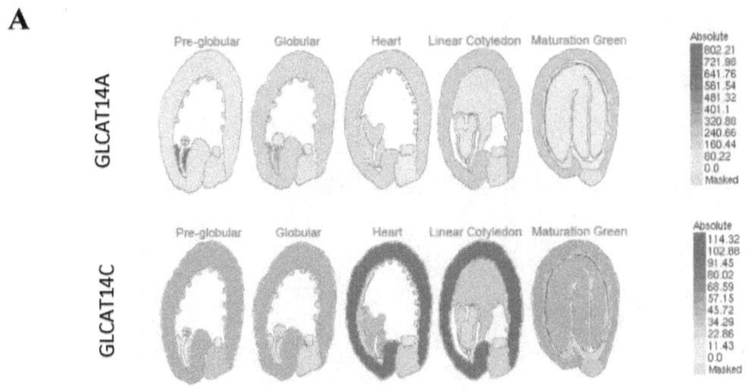

B

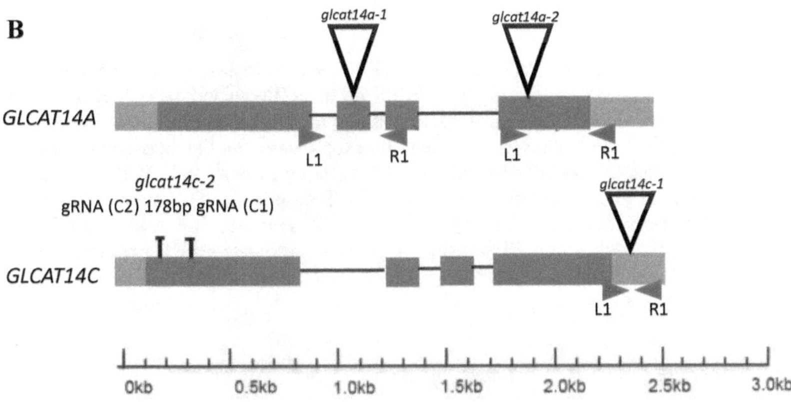

Figure 3-1: continued

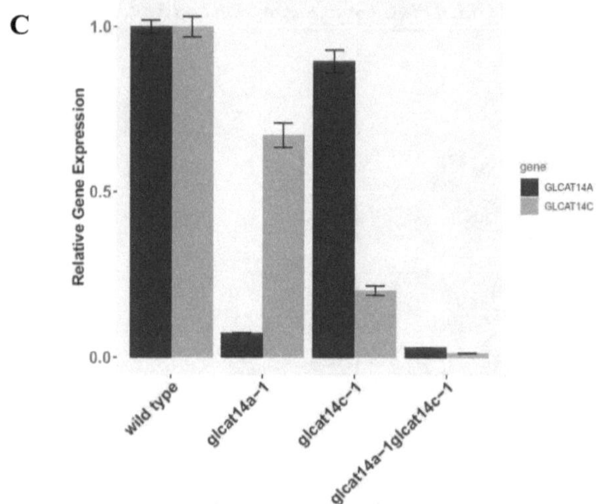

Note. A, *ATGLCAT14A* and *ATGLCAT14C* are expressed in the seed coat. B, T-DNA insertions (inverted triangles), CRISPR mutants (lines) and qRT-PCR primers (red arrowheads) are indicated. Also, the orange rectangles represent the promoters and the UTRs, the blue rectangles represent the exons while the line between the exons represents the introns. C, *GLCAT14A* and *GLCAT14C* gene expression in WT and *glcat14* mutant siliques at linear cotyledon stage (8 DAP). Transcript levels were normalized to the mean of one reference gene, the Arabidopsis actin 2 gene, *AtACT2*. Averages of three biological replicates ± SE are shown. Asterisks indicate significant differences compared with WT (Student's *t* test, P <0.01).

Determination of Mucilage Content

Three independent samples of 100mg seeds of WT, single mutants and double mutants were precisely weighed and extracted by vigorously shaking in 1 mL of distilled water for 5 min to isolate the non-adherent mucilage. The supernatants were completely transferred to separate tubes. One mL of distilled water was added to the remaining seeds and treated ultrasonically for 20 seconds (Zhao et al., 2017) at room temperature using a

Sonic Dismembrator Model 100 with the probe intensity set to 1. Supernatants were transferred to Eppendorf tubes to form the non-adherent mucilage. Both the non-adherent and the adherent mucilage contents were freeze dried and weighed to determine the mucilage content.

Microscopy and Image Analysis

Ruthenium Red Staining and Quantification of Mucilage Area. Mature dry seeds of WT and mutants were hydrated in distilled water, 50mM $CaCl_2$, 50mM EDTA, pH 8.0 and 1 M Na_2CO_3 for 30 mins, washed with water and then stained with the ruthenium red (RR) dye for 30 min at room temperature using 0.01% RR (Sigma, St Louis, MO, USA) as described elsewhere (Willats et al., 2001). Mature seeds prehydrated in distilled water was stained using 25 µg/ml fluorescent brightener 28 (Sigma) for, 20 min at room temperature as previously described (Willats et al., 2001). In both cases, seeds were shaken with a rotator and ruthenium red stained seeds were photographed using a Nikon SMZ1500 stereomicroscope coupled with a CCD Infinity 2 camera, while calcofluor stained seeds were imaged using a Zeiss LSM 510 confocal microscope. Pontamine staining of mature seeds hydrated in water and 50mM EDTA were carried out as described earlier (Anderson et al., 2010) using 0.01% pontamine fast scarlet S4B (Sigma) in 50mM NaCl for 30 min. Seeds were then de-stained four times with water before examination using a confocal microscope. Dry mature seeds were dropped in a 12-well plate containing 0.01% Ruthenium red stain without shaking and after shaking very briefly, and images were acquired using a light microscope. The ruthenium red-stained mucilage area was quantified using FIJI (ImageJ) as described previously (Voiniciuc et al., 2015). Regions of interest (ROI) were segmented in Fiji, and areas for the ROI were measured using the Analyze Particles function. Mucilage area was obtained by subtracting Seed area from Seed + Mucilage. Evaluation of statistical significance was conducted using R program by Tukey–Kramer HSD (P < 0.05).

Immunohistochemistry. Whole-seed immunolabeling was conducted according to a published method, except that seeds were shaken in water before immunolabeling and that seeds were stained with S4B after immunolabeling (Harpaz-Saad et al., 2008). Briefly, mature dry seeds were shaken in phosphate-buffered saline (PBS), pH 7.4 for 1 h. The supernatant (containing soluble mucilage components) was removed, and the remaining seeds with tightly bound mucilage were processed for immuno-fluorescence as follows: Seeds were shaken in 5% BSA in PBS for 30 min, washed with PBS, and incubated with the primary antibody CCRC-M35 (Pattathil et al., 2010) diluted 1/10 in 1% BSA in PBS for 1.5 h. Samples treated without a primary antibody served as a negative control. The specificities of the primary antibodies JIM5, JIM7, JIM13 and CCRC-M35 (CarboSource) have been extensively described (Pattathil et al., 2010). CBM3a, mostly specific to crystalline cellulose, and CBM28, mostly specific to amorphous cellulose regions (Blake et al., 2006), were treated as primary antibodies in identical solutions before treatment with mouse anti-histidine (Qiagen). Goat anti-rat secondary antibody conjugated to AlexaFluor488 was used against JIM5, JIM7 and JIM13, whereas goat anti-mouse conjugated to AlexaFluor488 (Molecular Probes; Invitrogen) was used as a secondary and tertiary antibody against the CCRC-M35 and CBMs, diluted 1/100 in 1% BSA in PBS for 1.5 h. Immunolabelled seeds were counterstained with S4B (Harpaz-Saad et al., 2012) and imaged using a Zeiss LSM 510 confocal microscope. Signal intensities for each antibody treatment were preserved across genotypes; however, the signal intensity was varied between treatments. Confocal micrographs were further processed using imageJ (Abramoff et al., 2004)

Enzyme-linked Immunosorbent Assay (ELISA) of CCRC-M35

Seeds (5 mg) of WT, single mutants and double mutants were precisely weighed and extracted by vigorously shaking in 1 mL of distilled water for 5 min to isolate the non-adherent mucilage. The supernatants were completely transferred to separate tubes. One mL of distilled water was added to the remaining seeds and treated ultrasonically for 20 seconds (Zhao et al., 2017) at room temperature using a Sonic Dismembrator Model 100 with the probe intensity set to 1. Supernatants were transferred to Eppendorf tubes to form the non-adherent mucilage. Two hundred (200 µl) of mucilage extracts were transferred to four wells on a 96-well ELISA plate (3598; Corning, Wiesbaden, Germany), while 200 µl of MilliQ water served as a negative control. The ELISA was carried out following methods described elsewhere (Yang et al., 2019), and the optical density (OD) value was read as the difference between the absorption value at 450 nm and 655 nm using a Synergy H1 microplate reader (BioTek, Bad Friedrichshall, Germany). The reading from each test well subtracted the value from the negative control well.

Dot Immunoblotting Assays

Non-adherent and adherent mucilage extracts (1mg/mL) were resuspended in water after freeze drying. A series of dilutions were prepared and a 1 µl aliquot was spotted onto a nitrocellulose membrane (Merck Millipore). After being air-dried, the membrane was blocked for 1h in 3% BSA in PBS, and then it was incubated for 1.5h in a 10-fold dilution of primary antibodies. After washing three times with PBS, membranes were incubated for 1.5h in horseradish peroxidase (HRP)-conjugated anti-rat (for Jim5 and Jim7) or anti-mouse (for CCRC M35) secondary antibodies in a 1000-fold dilution in in 1% BSA in PBS. Membranes were washed prior to color development in substrate solution [25mL de-ionized water, 5mL methanol containing 10mg mL-1, 4-chloro-1-naphthol and 30 µl 6% (v/v) H_2O_2]. After incubation for 30min at room temperature, the blots were rinsed with de-ionized water and photographed.

Scanning Electron Microscopy

Seed coat morphology was investigated using a JEOL JSM-6390 scanning electron microscope (Hitachi High-Technologies). Seeds were mounted on aluminum stubs using double adhesive tapestubs and sputter coated with a palladium alloy using a Cressington 208C high-resolution sputter coater (Ted Pella Inc.) at the Institute for Corrosion and Multiphase Technology, Ohio University. Electron micrographs were processed and measured using imageJ (Abramoff et al., 2004).

Determination of Monosaccharide Composition by HPAEC and Total Sugar Content

Non-adherent and adherent mucilage and whole seed alcohol insoluble residue (AIR) extracts were carried out as described previously (Zhao et al., 2017). One hundred microliters of mucilage extracts (adherent and non-adherent) and 50µl of 10mg/mL AIR were transferred to glass tubes and were hydrolyzed using 2 N trifluoroacetic acid (TFA) at 121°C for 90 min. TFA was removed by evaporation with N_2 gas. Samples were

dissolved in 500 µL milli-Q water containing 0.2mM cellobiose as an internal standard.
A standard sugar mixture (fucose, rhamnose, arabinose, galactose, glucose, xylose,
mannose, galacturonic acid, and glucuronic acid) was used for making the standard
curve. Monosaccharide compositions were calculated as molar percentages (mol %) and
in absolute amounts (mg/mg of seeds). All samples and standards were subjected to high
pH anion-exchange chromatography with pulsed amperometric detection (HPAEC-PAD)
using a Dionex PA-20 column (Thermo Fisher Scientific, Sunnyvale, CA, USA)
essentially as described here (Øbro et al., 2004). Total sugar (µg/mg seed) was
determined by phenol-sulfuric assay (Dubois et al., 1956) following sequential extraction
with 0.2% ammonium oxalate, 0.2N and then 2N sodium hydroxide for 1 h each with
vigorous shaking at 37°C.

Crystalline Cellulose Observation and Determination

For determination of crystalline cellulose content, 1 mL of distilled water was
added to 10 mg of mature dry seeds and treated ultrasonically for 20 seconds as described
previously (Zhao et al., 2017) at room temperature. The supernatant was transferred into
a separate tube, and the de-mucilaged seeds were kept for further analysis.
Approximately, ten (10) milligrams of seeds (exact weight recorded) alongside the de-
mucilage seeds were milled using steel balls for 5min. AIR from de-mucilaged and whole
seeds were isolated by two sequential washes with 1 mL of 70% (v/v) ethanol and
centrifugation for 10 min at 13,200g. After washing the AIR extract with 1:1 (v/v)
chloroform:methanol, followed by acetone, the pellet was dried for 5 min at 60°C.
Crystalline cellulose content was then determined as described previously (Updegraff,

1969), with minor modifications. The 2 mg of dry AIR (from whole and de-mucilaged seeds) together with 500µl of the total mucilage extracted previously were mixed with 1 mL of Updegraff reagent (acetic acid:nitric acid:water, 8:1:2 [v/v]) before incubation at 100°C for 30 min (Updegraff, 1969). After hydrolysis, the Updegraff-resistant pellet (containing only crystalline cellulose) was rinsed once with water, once with acetone, dried, and then hydrolyzed using 200 µL of 72% (v/v) sulfuric acid. Crystalline cellulose amounts were quantified colorimetrically at 620 nm in a spectrophotometer using the anthrone reagent (Updegraff, 1969). Seeds were also mounted in water on a microscope slide and observed with an epiflorescent microscope equipped with polarizing filters for birefringence by any crystalline cellulose in the investigated genotypes.

Uronic acid Estimation and Biochemical Determination of the Calcium and DM Content of HG

Whole mucilage was extracted by shaking 20 mg mature dry seeds in 500 µL of distilled water using an ultrasonication treatment as described previously. For the degree of methylesterification (DM), 200µL of supernatant was transferred into a new tube and saponified with 0.25 M NaOH for 1 h at room temperature with tube rotation. The reaction was neutralized with 0.25 M HCl (to give a total volume of 600µL) and centrifuged for 10 min at 10,000 g. The amount of methanol released after the saponification reaction was measured by a colorimetric method (Klavons and Bennett, 1986). Five hundred microliters of the supernatant was transferred into a new 1.5 mL tube, oxidized with 0.5 units of alcohol oxidase (Sigma-Aldrich) for 15 min at 25°C, and incubated with 500 µL of freshly prepared 0.02 M 2,4-pentanedione (dissolved in 2 M ammonium acetate and 0.05 M acetic acid) for 15 min at 60°C in a 1 mL total volume. After cooling on ice for 2 min, the absorbance was measured at 412 nm and quantified using a methanol standard curve.

The uronic acid content was determined by the meta-hydroxydiphenyl method (Blumenkrantz and Asboe-Hansen, 1973) using GalA as the standard. One hundred μL of the saponified mucilage solution was transferred into a new 1.5 mL microcentrifuge tube, and hydrolysed with 1.2 mL of concentrated sulfuric acid containing 0.0125M sodium tetraborate (Sigma-Aldrich) for 5 min at 100°C. After samples were cooled on ice, 25 μL of 0.15% (w/v) meta-hydroxydiphenyl (Sigma-Aldrich) in 0.5% (w/v) NaOH was added and absorbance was measured at 525 nm. The DM of HG in the mucilage extracts were calculated as the percentage molar ratio of methanol to uronic acid (Ralet et al., 2016). Uronic acid content of de-mucilaged and whole seeds were also estimated from AIR samples using methods described previously (Voiniciuc et al., 2013). For the estimation of calcium content, a commercial calcium colormetric assay kit (MAK022, Sigma-Aldrich, St. Louis, MO, USA) was used for calcium measurement following the manufacturer's protocol. Following mucilage extractions, calcium ions from the total mucilage extracts form a complex with the o-cresolphthalein in the assay kit, resulting in a color change from transparent to pink. The amount of calcium was determined using a UV spectrometer at OD_{575} and a standard curve made with different concentrations of $CaCl_2$ and values were expressed as percentages.

Results

Phylogenetic, Mutant Characterization and Gene Expression Analyses of the

GLCAT14A and GLCAT14C Genes

GLCATs are involved in the transfer of GlcA to Type II AG glycans. Although eleven confirmed and/or putative β-GLCATs have been identified in Arabidopsis, phylogenetic analysis showed that *GLCAT14A* and *GLCAT14B* appear to be paralogs, while *GLCAT14C* is phylogenetically distinct from *GLCAT14A* and *GLCAT14B* (**Figure 3-2**). Seed microarray data displayed by the eFP browser (Winter et al., 2007; Le et al., 2010) revealed that *GLCAT14A* and *GLCAT14C* had elevated expression in the seed coat (**Supplemental Figure 3-1**) and in seed development, primarily during the heart and linear cotyledon stages (**Figure 3-1A**). To this end, we examined *glcat14a-1* and *glcat14c-1* single mutants and a *glcat14a-1glcat14c-1* double mutant to reveal the role of these genes in seed mucilage biosynthesis.

Figure 3-2

Phylogenetic Tree of CAZy GT14 Proteins from Eight Species

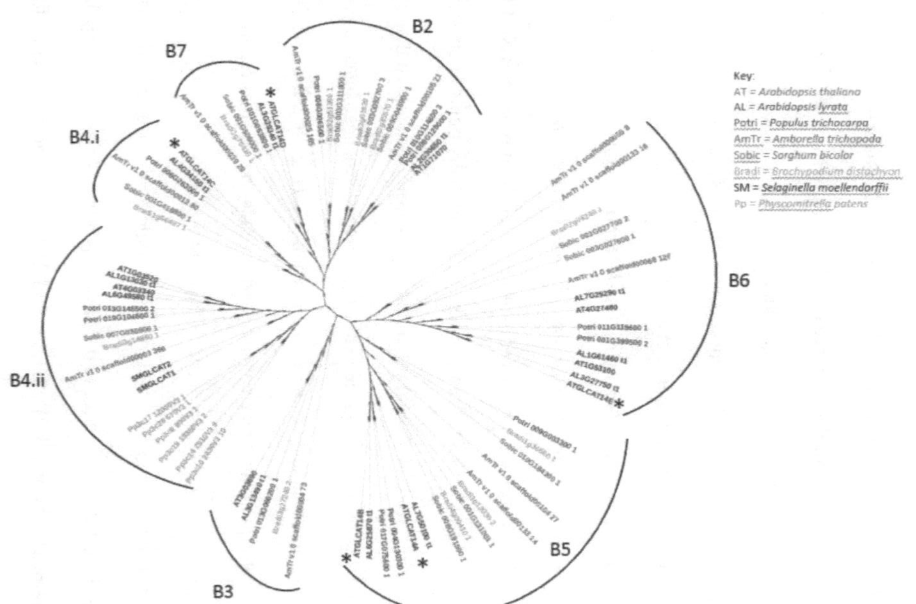

Note. Five GT14s (denoted by an asterisk) have previously shown GlcA transferase (GLCAT) activity. The species protein sequences were used to construct the phylogeny (See Materials and Methods) and are indicated in the figure legend. ATGLCAT14A belongs to the B5 clade while ATGLCAT14C belongs to B4.i clade.

Single mutants (*glcat14a-1* and *glcat14c-1*) and the double mutant (*glcat14a-1glcat14c-1*) were examined for the expression of *GLCAT14A* and *GLCAT14C* using quantitative reverse transcription (qRT)-PCR. Given the expression of *GLCAT14A* and *GLCAT14C* across seed developmental stages (**Figure 3-1A**), we examined their expression at the linear cotyledon stage (8 DAP) in WT, *glcat14a-1*, *glcat14c-1* and *glcat14a-1glcat14c-1* mutants and observed a significant reduction in gene expression of *GLCAT14A* and *GLCAT14C* in both the single and double mutants (**Figure 3-1C**). We were unable to confirm the presence of a T-DNA insertion in the SALK_051810 line for *glcat14c-2* but a CRISPR knockout of the *GLCAT14C* gene close to its 5' end resulted in a 178 bp gene deletion, and produced similar phenotypes as the *glcat14c-1* (SALK_005705) mutant (Zhang, et al., 2020).

The glcat14a-1 and glcat14a-1glcat14c-1 Mutants Have Distinct Seed Coat Mucilage Phenotypes in Response to Different Chemical Extractants

WT and mutant seeds were hydrated in distilled water and Na_2CO_3 and stained with ruthenium red (RR), a red dye which preferentially binds to unesterified pectin (Hou et al., 1999). Seeds shaken in water and stained with RR showed that *glcat14a-1* seeds had a smaller mucilage capsule, while the adherent mucilage layer in *glcat14a-1glcat14c-1* mutant seeds was undetectable compared to WT (**Figure 3-3A**). Similarly, the quantification of mucilage areas in hydrated seeds showed that relative to WT, *glcat14a-1* and *glcat14c-1* had a 57.5% and 2.7% reduction in mucilage area, respectively, while the mucilage area in *glcat14a-1glcat14c-1* could not be determined (**Figure 3-3D**). Given the loss of adherent mucilage in the *glcat14a-1glcat14c-1* seeds (**Appendix B: Supplemental Figure 3-2D, 2H, 2L and 2N**), it was unclear whether this mucilage deficient phenotype was due to mucilage extrusion defects or to the repartitioning of the mucilage layers. To answer this question, we investigated how the mutants extrude mucilage by dropping mature dry seeds in 0.01% RR dye. Results showed that *glcat14a-1glcat14c-1* seeds extruded mucilage like WT and single mutants, but then began "peeling off" the mucilage upon gentle shaking (**Figure 3-3C**). While hydrating the seeds in water, the vast majority of the *glcat14a-1glcat14c-1* seeds floated (**Figure 3-4D**), even after 1hr of extended contact with water. Also, the double mutant seeds were packed together (**Figure 3-4H**), remained afloat and even germinated after 48hrs (**Figure 3-4L**).

Chemical extraction with Na_2CO_3 extracts pectins by cleavage of cross-linking ester linkages (Selvendran and Ryden, 1990; Fry, 2000; McCartney and Knox, 2002). Treatment of the WT and single mutant seeds with 1M Na_2CO_3 resulted in the rupturing of the cell wall to form organized 'pyramidal' arrangements of primary cell wall remnants attached to the columella, which was visualized as dark staining points on the seed surface. In *glcat14a-1glcat14c-1* seeds, the tangential and/or radial cell wall appears to be intact, lacking both the 'pyramidal structure' and the adherent mucilage (**Figure 3-3B**).

Figure 3-3

The glcat14a-1 and glcat14a-1glcat14c-1 Mutants have Seed Mucilage Defects

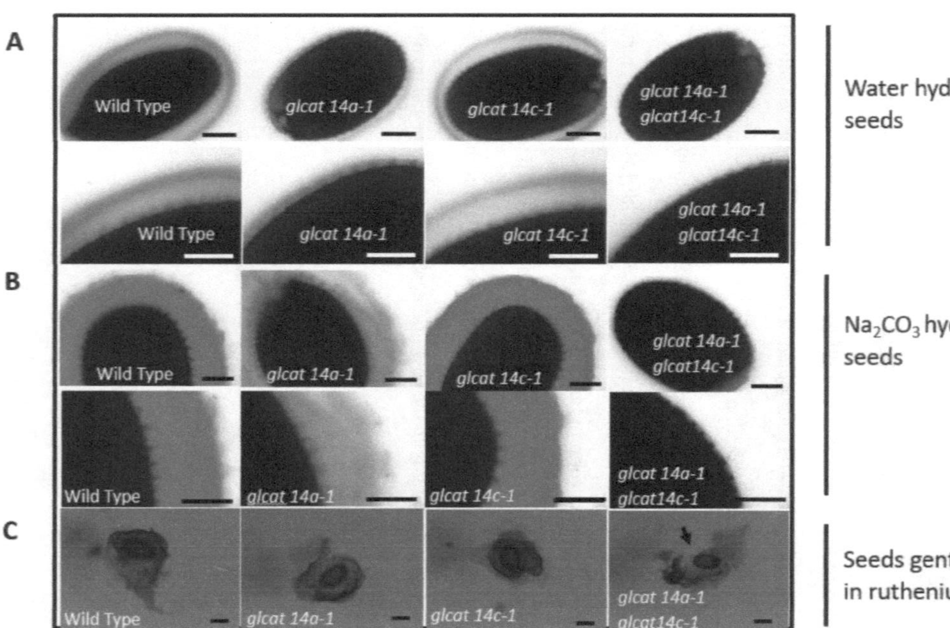

Figure 3-3: continued

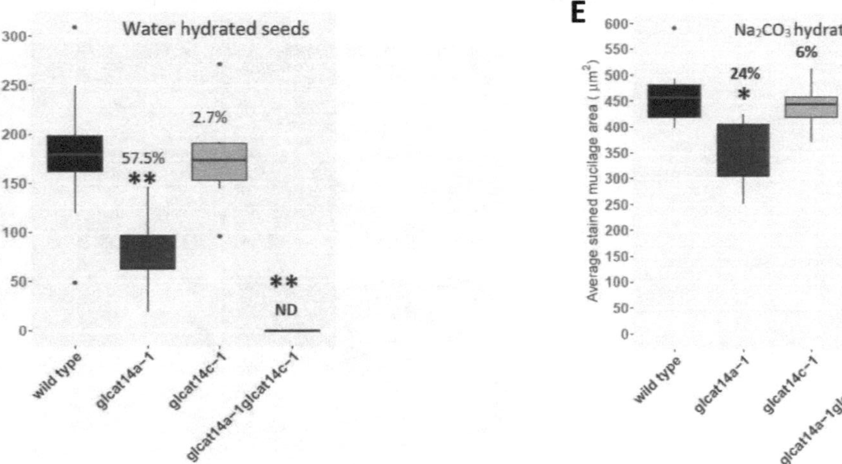

Note. A, Seeds hydrated in water were stained with 0.01% RR. B, Seeds hydrated in 1M Na_2CO_3 and stained with 0.01% RR. Loss of adherent mucilage was observed in the *glcat14a-1glcat14c-1* double mutant seeds coupled with the "peeling off" of the adherent layer from the seed coat when gently shaken, as indicated by an arrow (C). Note that in the WT and *glcat14* single mutants, the adherent layer surrounded the seed coat. Quantification of the average stained mucilage area for water hydrated seeds (D) and Na_2CO_3 hydrated seeds (E). Box plots were generated from three biological replicates of (>20 seeds each). The percentage (%) decrease in mucilage area (D and E) were indicated for each mutant relative to the WT. The single and double asterisk marks a significant decrease compared with WT (Student's *t*-test, P < 0.05 for single asterisks and P < 0.01 for double asterisks). ND- Not detected. Bars = 150μm

Figure 3-4

Seed Floating and Compactibility Were Displayed in glcat14a-1glcat14c-1 Mutants

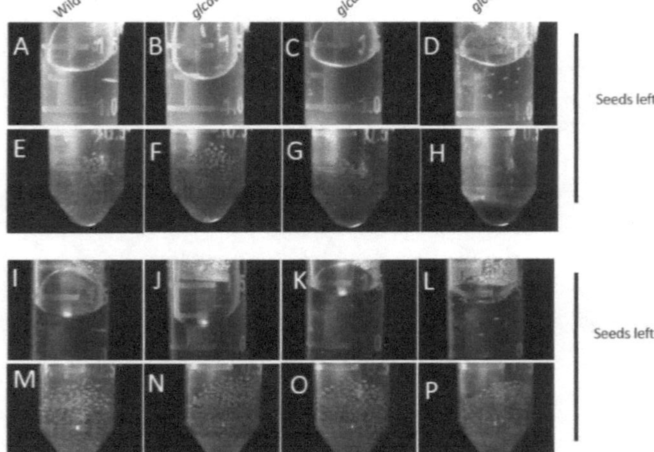

Note. Seeds of WT and mutants were shaken and left to stand for one hour (A-F) and 48 hours (I-P). A-D and I-L represents the top layer of the sample tubes while E-F and M-P represents the lower part of the sample tubes. *glcat14a-1glcat14c-1* mutant seeds floated (D) and compacted (H) after being left to stand for 1 hr and germinated after 48 hr while staying afloat. Similar seed quantities were added to each tube. Bar = 0.75mm

Similarly, the RR dye staining intensity of adherent mucilage was lower in *glcat14a-1* compared to WT, while the staining in *glcat14c-1* was indistinguishable from WT. Quantification of mucilage areas in mature seeds hydrated in 1M Na_2CO_3 revealed that *glcat14a-1* and *glcat14c-1* mutants had a 24% and 6% reduction in mucilage area, respectively, while mucilage area of *glcat14a-1glcat14c-1* seeds could not be determined due to the significant loss of adherent mucilage in the seed coat (**Appendix B: Supplemental Figure 3-2E-2H and Figure 3-3E**).

GLCAT14A and GLCAT14C Influences Cellulose Ray Morphology and Cellulose Deposition

WT and *glcat14* mutant seeds were hydrated in distilled water and 50mM EDTA and were examined for the precise distribution of cellulose in the mucilage capsule using the S4B dye, which binds cellulose (Anderson et al., 2010). Results showed that WT and single mutant seed mucilage capsules displayed ordered and intense S4B-labeled cellulosic rays that projected outwards from the top of the columellae, as well as diffuse S4B signals between rays following water (**Figure 3-5A, Upper panel**) and EDTA extractions (**Figure 3-5A, Lower panel**). In contrast, *glcat14a-1glcat14c-1* seeds were characterized by irregular cellulose ray organization with incompletely detached primary cell walls in water hydrated seeds, and primary cell wall remnants bound tightly to the periphery of the extruded mucilage for EDTA hydrated seeds (**Figure 3-5A**). To further characterize the fine structure and distribution of cellulose in seed adherent mucilage, we used calcofluor, a dye which binds β-glucans (Willats et al., 2001), and two carbohydrate-binding modules (CBMs; CBM3a and CBM28) immunolabelled in parallel with the S4B stain. CBM3a binds preferentially to crystalline cellulose structures, whereas CBM28 binds preferentially to amorphous cellulose structures (Blake et al., 2006). Similar to the RR staining, we observed the loss of the feathery ray structure of the calcofluor stained adherent mucilage layer in the *glcat14a-1glcat14c-1* seeds compared to the WT (**Figure 3-5B and Appendix B: Supplemental Figure 3-2N**). In the WT and single mutants, CBM3a displayed a mustache tip-like structure that was concentrated especially at the outer periphery, whereas S4B stained the inner adherent layer and the rays above the columella. In contrast to WT, *glcat14a-1glcat14c-1* seeds had more severe defects as indicated by the absence of S4B stained ray-like structures, with some CBM3a immunolabelling detected at regions closest to the seed coat (**Figure 3-5C and Appendix B: Supplemental Figure 3-3A**). CBM28 labeling of *glcat14a-1* and *glcat14c-1* had a similar pattern as the WT but with reduced intensity; whereas, in *glcat14a-1glcat14c-1* double mutant, mucilage labeling was almost completely absent (**Figure 3-5D**). Similarly, the adherent mucilage was observed for birefringence by any crystalline cellulose present and results indicated that WT and single mutant seeds showed bright regions with visible rays of crystalline cellulose within the adherent mucilage, but such birefringence was absent in *glcat14a-1glcat14c-1* seeds, except for the bright spots on the edges of seeds (**Figure 3-5E**). Similarly, crystalline cellulose content in total mucilage, demucilaged and whole seeds showed that *glcat14a-1glcat14c-1* mutants had significantly reduced crystalline cellulose content relative to the WT (**Figure 3-5F**).

Figure 3-5

Cellulose Deposition is Altered in glcat14a-1 glcat14c-1 Double Mutants

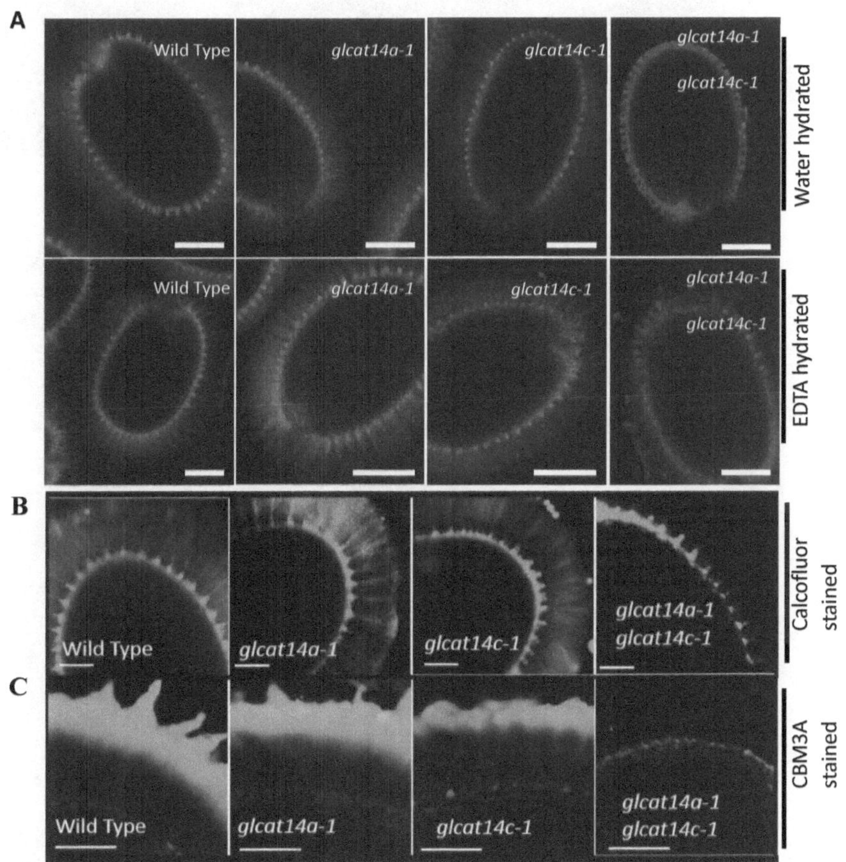

Figure 3-5: continued

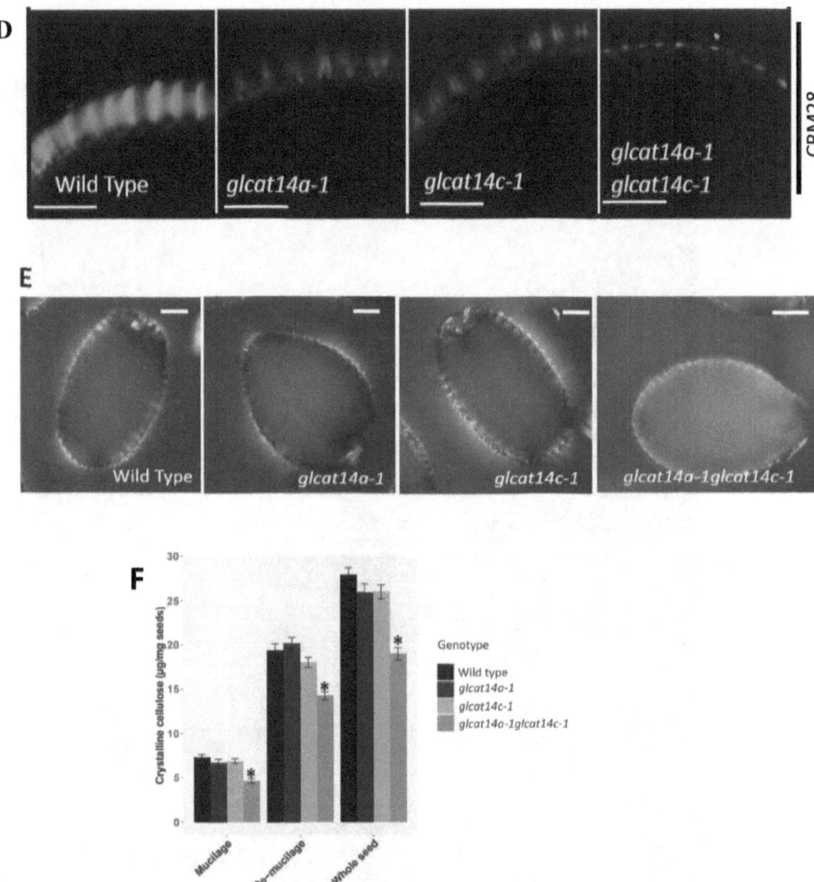

Note. A, Pontamine fast scarlet (S4B) cellulose staining of the adherent mucilage of water hydrated mature seeds (Upper panel) and EDTA hydrated seeds (lower panel). B, Calcofluor staining of the adherent mucilage of mature seeds. Immunolabelling of mature seeds with CBM3a (C) and CBM28 (D) counterstained with the S4B dye. Visualization of polarized light birefringence by crystalline cellulose in adherent mucilage released from WT and mutant seeds (E). Quantification of crystalline cellulose contents in whole seeds, demucilaged seeds, and in the mucilage of WT and mutants (F) using the Updegraff assay. Bars = 100µm

Mucilage Pectin Components Altered in glcat14a-1 and Highly Altered in glcat14a-1glcat14c-1

In addition to hydrating matured seeds in water and Na_2CO_3 and staining with RR, WT and *glcat14* mutant seeds were shaken in 50mM EDTA to investigate whether there is any residual mucilage trapped in the seed coat. Typically, cation chelators like EDTA can facilitate mucilage extrusion by disrupting crosslinks in unesterified HG chains (Rautengarten et al., 2008; Arsovski et al., 2010). Hydration of mature seeds in 50mM EDTA, pH 8.0 showed that in contrast to the WT and two single mutants, *glcat14a-1glcat14c-1* double mutant seeds had primary cell wall remnants attached to the seed coat coupled with loss of adherent mucilage (**Appendix B: Supplemental Figure 3-4D and 4H**). Given the reported role of glucuronic acid in calcium binding (Lopez-Hernandez et al., 2020; Lamport and Varnai, 2013), we investigated whether the addition of calcium ions impact the pectic gel matrix of the adherent mucilage in *glcat14* mutant seeds. While the intensity of the RR stained mucilage of the *glcat14a-1* and *glcat14c-1* seeds shaken in 50mM $CaCl_2$ were comparable to the WT (**Appendix B: Supplemental Figure 3-4I-4K, 4M-4O**), the RR staining intensity of *glcat14a-1glcat14c-1* adherent mucilage still displayed loss of adherent mucilage (**Appendix B: Supplemental Figure 3-4L, 4P and 4Q**).

Three pectin antibodies, JIM5 and JIM7, and CCRC-M35, were used in conjunction with S4B staining to examine the distribution of pectin relative to cellulose in the adherent mucilage. JIM5 and JIM7 are specific for partially methylesterified (up to 40%) and methyl esterified (up to 80%) HG respectively (Varidenbosch et al., 1989;

Knox et al., 1990), whereas CCRC-M35 recognizes unsubstituted RG-I backbones present in Arabidopsis seed mucilage (Young et al., 2008; Arsovski et al., 2010; Pattathil et al., 2010). CCRC-M35 labeling of WT and *glcat14* single mutant seeds appeared to surround the ray structures at the periphery of the mucilage halo (**Appendix B: Supplemental Figure 3-5A1-L1**), whereas in the *glcat14a-1glcat14c-1* seeds, the CCRC-M35 labeling appeared to be at the surface of the seed coat and was not concentrated in a ray-like manner (**Figure 3-6A, 6D and Appendix B: Supplemental Figure 3-5, panel J1-L1**). Similarly, the distribution of partially methylesterified HG was also examined using the JIM5 antibody. Surprisingly, the diffuse JIM5 staining between columella present in the WT was absent in *glcat14a-1* and reduced in *glcat14c-1* mutant seeds (**Appendix B: Supplemental Figure 3-5, panel A2-I2**). In contrast to WT, the *glcat14a-1glcat14c-1* seeds were intensely labeled at regions close to the columella, and at regions that appear to be incompletely detached primary cell wall fragments (**Figure 3-6B, 6E and Appendix B: Supplemental Figure 3-5, panel J2-L2**). Similar observations were made with JIM7 labelling with intense staining observed around the columella regions for *glcat14a-1* and *glcat14a-1glcat14c-1* (**Figure 3-6C; Appendix B: Supplemental Figure 3-5, panel D3-F3 and J3-L3**). To exclude the possibility that the changes in the HG esterification resulted from increased epitope accessibility, the calcium content and the degree of methylation (DM) of HG in the total mucilage extracts were determined using biochemical assays. Relative to WT, the calcium content decreased by 4.5%, 5% and 37.5% in *glcat14a-1*, *glcat14c-1* and *glcat14a-1glcat14c-1* mutants, respectively, while the DM of HG increased by 47%, 32% and 5.3% in *glcat14a-1*, *glcat14c-1* and *glcat14a-1glcat14c-1* mutants, respectively (**Figure 3-6F**). Similarly, in contrast to WT, the uronic acid content of the non-adherent mucilage increased significantly in both the single and double mutants while a significant reduction was observed in the *glcat14a-1glcat14c-1* mutant in the adherent layer (**Figure 3-6G**). The mucilage polymers were further assessed using immunoblot analyses and showed differences in polymer constituents between the WT and *glcat14* mutants (**Appendix B: Supplemental Figure 3-6A-6C**). While we observed increased CCRC-M35 epitope binding for *glcat14a-1* and *glcat14c-1* in the adherent mucilage, a significant reduction was observed in *glcat14a-1glcat14c-1* mutants relative to WT (**Appendix B: Supplemental Figure 3-6D**). Notably, we found that JIM13 epitopes were detected and localized to the columella, but we did not observe any difference in the JIM13 signal between WT and *glcat14* mutants (**Appendix B: Supplemental Figure 3-5, panels A4, D4, G4 and K4**). Overall, our results obtained using biochemical and immunolabelling approaches provide evidence that the pectic organization in the seed coat mucilage is severely affected in *glcat14a-1glcat14c-1* mutants.

Figure 3-6

Pectin Deposition is Altered in glcat14a-1 Single Mutants and glcat14a-1 glcat14c-1

Double Mutants

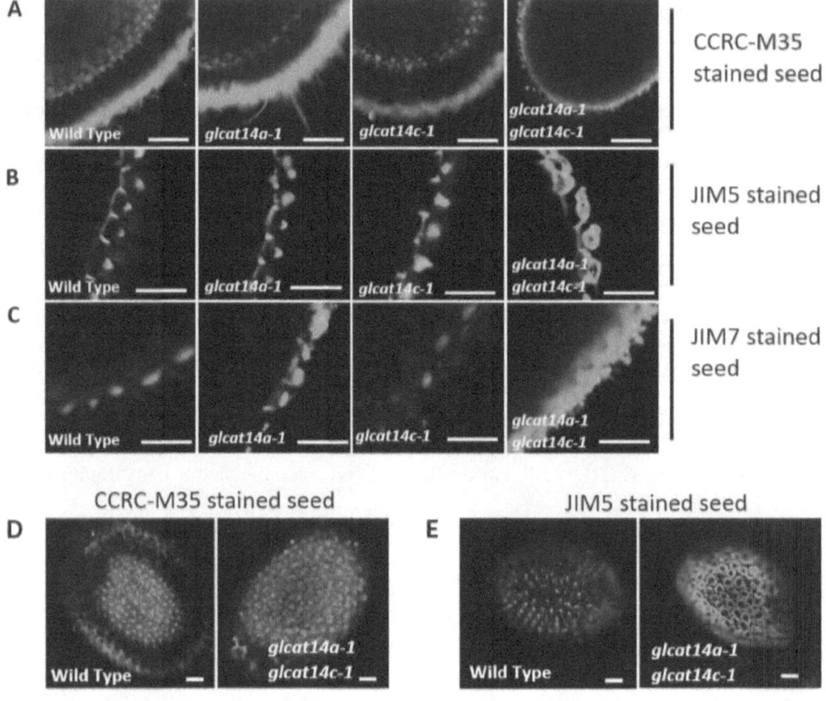

Figure 3-6: continued

F

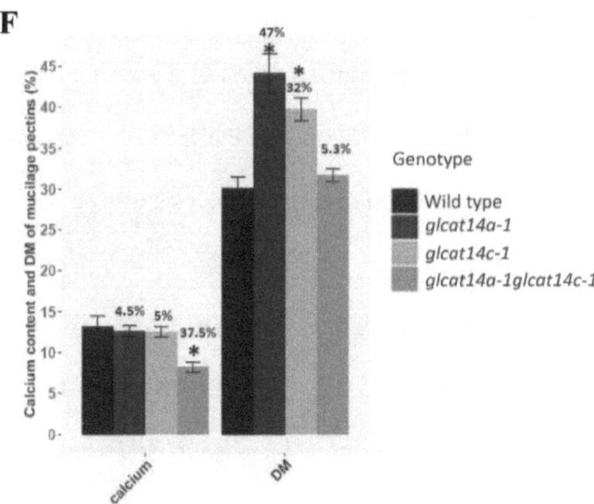

G

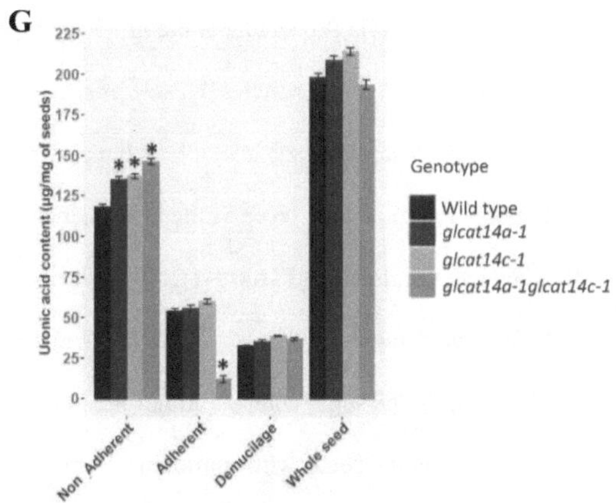

Note. Immunolabelling with CCRC-M35 (A, D), JIM5 (B, E) and JIM7 (C) of the adherent mucilage and counterstained with S4B. Biochemical determination of calcium content and degree of pectin methylesterification expressed as a percentage (F). Uronic acid (G) was estimated as described (see materials and methods). Percentage (%) increase or decrease relative to WT is indicated in F. Values represents the means ± SD of four biological replicates. The single asterisk indicates significant differences compared to WT (Student's *t*-test, P < 0.05 for single asterisks). Bars = 100µm; DM-degree of methylesterification

Sugar Distribution Altered in glcat14a-1glcat14c-1 Mucilage

The two most abundant sugars of the mucilage carbohydrates are rhamnose and galacturonic acid; together they make up approximately 80% of the total mucilage (Western et al., 2000; Griffiths et al., 2014). To examine the effects of *glcat14a-1glcat14c-1* mutation had on mucilage composition, sugar analysis was performed. We observed a significant reduction in GlcA content (mol%) for *glcat14a-1glcat14c-1* mutants in the non-adherent mucilage relative to WT. Similarly, GlcA was not detected in the adherent mucilage of the *glcat14* mutants, only in the WT (**Table 3-1**). Although, we observed a slight increase in Gal and Xyl for *glcat14a-1glcat14c-1* seeds, these increases were not significant (P>0.05). Compared to WT, *glcat14a-1* and *glcat14c-1* had significant alterations in the galacturonic acid (GalA) content in the mucilage layers, while the remaining sugars were comparatively similar to the WT. Surprisingly, the *glcat14a-1glcat14c-1* mutant showed significant increase in GalA with a corresponding decrease in other sugars in the non-adherent mucilage layer, while in the adherent layer, we observed an increase in other sugars except GalA (**Table 3-1**). Also, we observed a significant reduction in the total mucilage content in *glcat14a-1glcat14c-1* relative to WT (**Figure 3-7**). To further confirm the shift in the sugar composition of the mucilage layers, sequential extractions of WT and mutant seeds with ammonium oxalate, 0.2 N NaOH and 2 N NaOH showed a significant increase in the total sugar content for the *glcat14a-1glcat14c-1* mutant in ammonium oxalate and 0.2 N NaOH extracts, and a significant decrease in the 2 N NaOH extracts (adherent layer) (**Table 3-2**).

Table 3-1

Monosaccharide Composition Analysis of WT and glcat14 Mutants

	Non-Adherent mucilage (mol %)				Adherent mucilage (mol%)				Whole Seed (mol%)			
Sugar	Wild Type	*glcat14a-1*	*glcat14c-1*	*glcat14a glcat14c*	Wild Type	*glcat14a-1*	*glcat14c-1*	*glcat14a glcat14c*	Wild Type	*glcat14a-1*	*glcat14c-1*	*glcat14a glcat14c*
Fuc	1.31 ± 0.06	1.12 ± 0.02	1.05 ± 0.03	**0.44 ± 0.01***	0.71 ± 0.04	0.82 ± 0.02	0.65 ± 0.04	**2.66 ± 0.01****	2.02 ± 0.13	1.92 ± 0.14	1.62 ± 0.53	1.79 ± 0.36
Rha	31.53 ± 0.51	28.61 ± 0.17	27.44 ± 0.41	**29.2 ± 0.39****	25.58 ± 0.43	27.85 ± 0.69	28.47 ± 0.41	**26.63 ± 0.22****	17.03 ± 0.25	18.42 ± 0.24	18.03 ± 0.27	16.67 ± 0.54
Ara	0.38 ± 0.01	0.34 ± 0.05	0.29 ± 0.08	**0.15 ± 0.01****	0.09 ± 0.012	0.12 ± 0.013	0.18 ± 0.015	**0.60 ± 0.01****	20.59 ± 0.13	19.37 ± 0.37	20.24 ± 0.67	19.61 ± 0.46
Gal	3.68 ± 0.09	3.59 ± 0.09	4.97 ± 0.08	**2.74 ± 0.05***	5.97 ± 0.02	6.09 ± 0.11	5.78 ± 0.04	**13.20 ± 0.07****	15.76 ± 0.14	15.39 ± 0.45	15.81 ± 0.26	16.77 ± 0.32
Glc	3.80 ± 0.01	3.92 ± 0.03	3.63 ± 0.02	**2.28 ± 0.06***	5.87 ± 0.07	5.98 ± 0.06	6.36 ± 0.05	**10.85 ± 0.06****	4.69 ± 0.12	5.16 ± 0.25	4.65 ± 0.12	**6.72 ± 0.36***
Xyl	7.95 ± 0.01	7.55 ± 0.02	5.53 ± 0.02	**5.4 ± 0.05***	4.8 ± 0.06	5.43 ± 0.22	4.55 ± 0.03	**7.66 ± 0.08****	12.76 ± 0.23	12.03 ± 0.66	12.39 ± 0.37	13.04 ± 0.83
Man	3.42 ± 0.04	3.03 ± 0.03	2.68 ± 0.04	**1.20 ± 0.03***	3.23 ± 0.25	2.89 ± 0.36	2.84 ± 0.17	**6.53 ± 0.11****	2.76 ± 0.25	2.11 ± 0.11	2.00 ± 0.39	2.38 ± 0.82
GalA	47.59 ± 0.13	51.55 ± 0.68	**54.15 ± 0.73***	**58.55 ± 0.92****	53.7 ± 0.35	**50.81 ± 0.67***	51.17 ± 0.98	**31.87 ± 0.11****	24,39 ± 0.16	25.60 ± 0.45	25.25 ± 0.61	23.02 ± 0.69
GlcA	0.34 ± 0.01	0.30 ± 0.03	0.26 ± 0.03	**0.04 ± 0.01****	0.05 ± 0.01	**n.d**	**n.d**	**n.d**				

Note. *P<0.05; **P<0.01; n.d Not detected; Sugars significantly different from WT are indicated in bold.

Mucilage Weights of WT and glcat14 Mutants

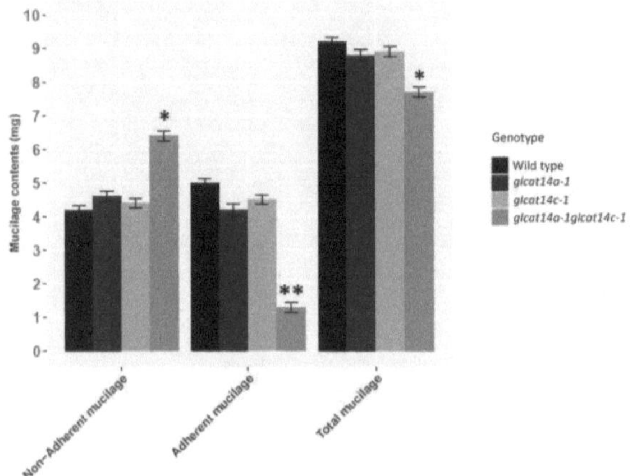

Note. The total mucilage is a combination of the adherent and the non-adherent mucilage. Significant alterations in mucilage content were observed in *glcat14a-1 glcat14c-1* double mutants in comparison to WT. Values represents the means ± SD of three biological replicates. The asterisk indicates significant differences compared to WT (Student's *t*-test, P < 0.05 for single asterisks, P < 0.01 for double asterisks).

Table 3-2

Total Sugar Estimation in WT and glcat14 Mutants

Extracts[a]	WT	*glcat14a-1*	*glcat14c-1*	*glcat14a-1glcat14c-1*
Ammonium oxalate	5.74 ± 0.46	6.03 ± 0.35	5.97 ± 0.12	**7.8 ± 0.22***
0.2N NaOH	11.73 ± 0.63	**13.9 ± 0.42***	12.94 ± 0.83	**15.04 ± 0.42***
2N NaOH	10.59 ± 0.72	9.6 ±0.32	10.23 ± 0.29	**4.39 ± 0.33***
Total Sugar	28.06	29.53	29.14	27.23

Note. [a]Quantification of total sugars from WT and *glcat14* mutants mucilage sequentially extracted using 0.2% ammonium oxalate, 0.2 N NaOH and 2 N NaOH neutralized and assayed with the phenol-sulfuric acid method against glucose standards. Results are given as ug/mg seed ± SE. Significant differences from WT, (P<0.05) are indicated with an asterisk and shown in bold.

GLCAT14A and GLCAT14C Required for Seed Coat Epidermal Cell Development

We employed SEM to observe any potential alterations in the SCE cells that are reflective of the changes in the cell wall polymer characteristics. While the surface morphology of the *glcat14a-1* and *glcat14c-1* seeds were indistinguishable from the WT (**Appendix B: Supplemental Figure 3-7**), the surface morphology of the *glcat14a-1glcat14c-1* seeds displayed alteration of the SCE cells characterized by morphostructural changes in the radial cell walls of the hexagonal plane (**Figure 3-8A, and 8B**), coupled with changes in the appearance of the columella after water imbibition (**Figure 3-8C, and 8D**). Similarly, in contrast to WT, we observed a significant increase in the columella area of *glcat14a-1glcat14c-1* seeds before and after shaking the seeds in water (**Figure 3-8E and 8F**). We also measured the seed size and found that *glcat14a-1glcat14c-1* seeds had significantly reduced seed length and width relative to WT (**Figure 3-9A, 9B and 9C**).

Figure 3-8

Columella and Radial Cell Wall are Severely Impaired in the glcat14a-1 glcat14c-1

Double Mutant Seeds

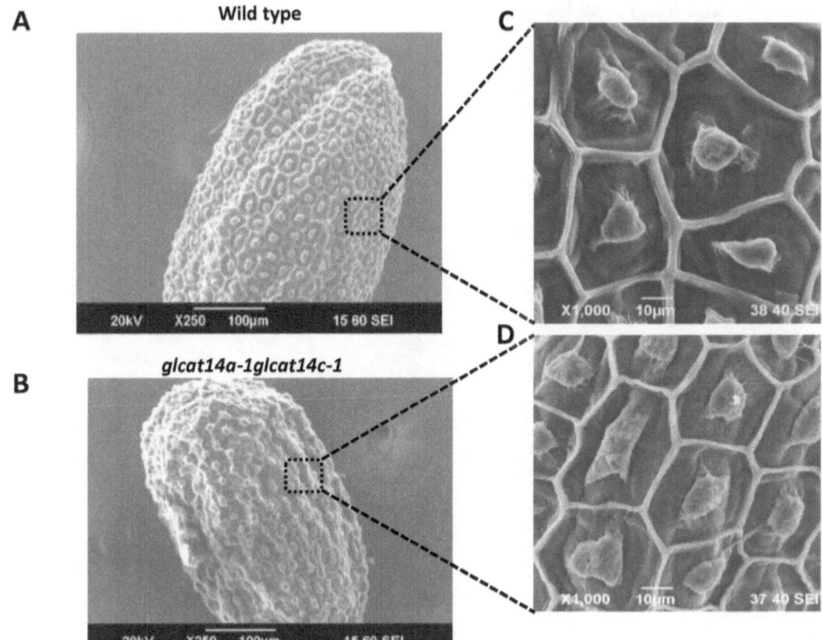

Figure 3-8: continued

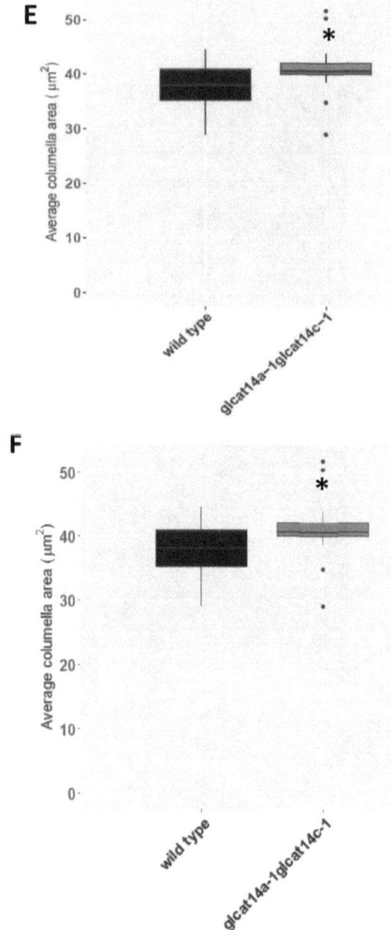

Note. Scanning electron microscopy of the seed coat surface of WT (A) and *glcat14a-1 glcat14c-1* double mutant seeds (B). Seeds were imbibed in water and visualized for alterations in the seed coat surface in WT (C) and *glcat14a-1 glcat14c-1* double mutant seeds (D). Quantification of columella area before water hydration (E) and after water hydration (F). *, Significant differences from WT (Student *t* test, P < 0.05).

Figure 3-9

Seed Size is Reduced in the glcat14a-1 glcat14c-1 Double Mutant Seeds

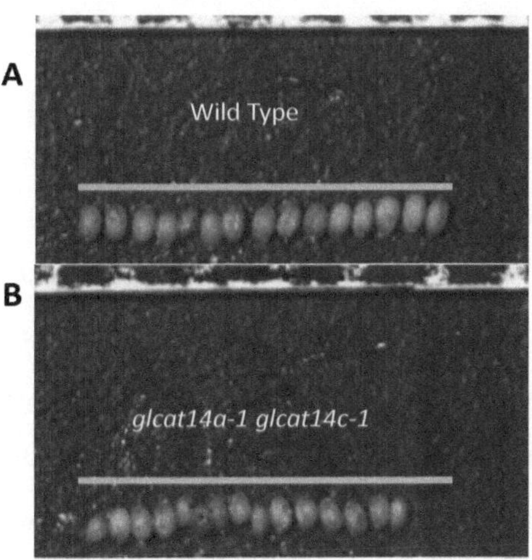

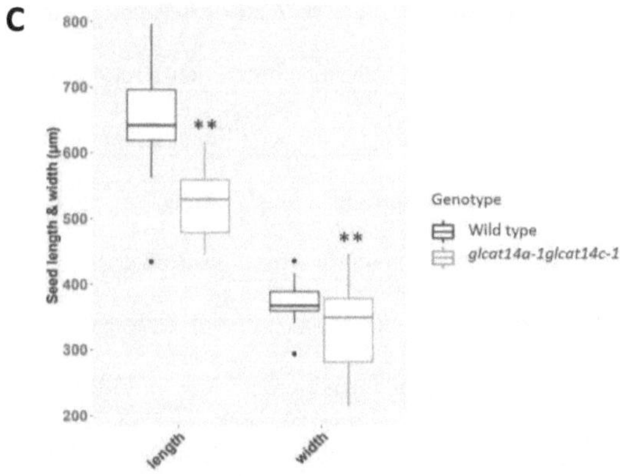

Note. A, WT seeds were compared to *glcat14a-1 glcat14c-1* double mutant seeds (B). Quantification of the length and width of WT and *glcat14a-1 glcat14c-1* double mutant seeds (C), n = 45 seeds. ** Significant differences from WT (Student *t* test, P < 0.01). Bar = 1mm; Blue and white boarders above panel A and B are the measuring ruler.

Discussion

Over the years, research efforts have been tailored towards identifying components of the mucilage polysaccharides that are important in the organization of the mucilage matrix. Despite the significant advances made in the discovery of the GTs involved in mucilage formation, accumulating evidence indicates that our understanding of the mucilage polysaccharide matrix formation and organization is far from complete. Here, we provide evidence that two *GLCAT* genes involved in the transfer of GlcA to type II AGs in AGPs are critically important in controlling the structural integrity of the mucilage polysaccharide matrix and seed size and elucidates the relationship between β-linked GlcA residues in AGPs and the stability of the mucilage polysaccharide architecture.

Loss of Function of GLCAT14A and GLCAT14C Results in Seed Flotation

Arabidopsis seeds, when imbibed in water, form a dense mucilage layer around the seed which makes them sink. Interestingly, the seed floating phenotype displayed by the *glcat14a-1 glcat14c-1* seeds phenocopies previously characterized mucilage mutants of the floating mucilage-releasing (FMR) natural Arabidopsis accessions. The genes responsible for the FMR defects were speculated to be involved in cellulose formation given the absence of cellulose labelling on the adherent layers in FMR mutant seeds (Saez-Aguayo et al., 2014). A similar seed floating phenotype was observed in *irx14* mutants with impaired xylan synthesis (Voiniciuc et al., 2015), suggesting that in addition to the reduction of cellulose, the absence of other mucilage polymers can also contribute to the FMR phenotype. Our result show that the *glcat14a-1 glcat14c-1* seeds displayed phenotypes similar to those demonstrated by *irx14* mutant (Voiniciuc et al., 2015) and FMR natural Arabidopsis accessions (Saez-Aguayo et al., 2014), and extends our previous knowledge on how intricately intertwined and interdependent the matrix polymers are, and how the genetic disruption of one of the interacting partners can have a profound effect on mucilage matrix architecture. It is worth noting that seed floating phenotypes have also resulted from impaired mucilage release (Arsovski et al., 2010; Saez-Aguayo et al., 2013), but that was not the case for the *glcat14a-1 glcat14c-1* seeds, which released mucilage upon contact with water-dissolved RR dye (**Figure 3-3C**). This seed floating phenotype appears to be evolutionarily advantageous in improving seed dispersal over rivers, while still retaining its germination properties (Voiniciuc et al., 2015).

Loss of Function of GLCAT14A and GLCAT14C Results in Severe Mucilage Phenotypes

Several studies investigating the loss of function of genes/enzymes involved in mucilage formation have been characterized by the repartitioning of the mucilage layers upon water imbibition. In water hydrated seeds, we observed that *glcat14a-1* seeds displayed a significant reduction in adherent mucilage relative to WT, with a loss of adherent mucilage in *glcat14a-1 glcat14c-1* seeds (**Figure 3-3A**). The distinct mucilage phenotypes of *glcat14a-1* and *glcat14c-1* seeds may reflect the distinct mechanisms of action in the glucuronidation process that may be influenced by the glycan architecture based on the finding that ATGLCAT14A prefers β-1,6- galactans while ATGLCAT14C prefers β-1,3-galactans as substrates in an *in-vitro* assay (Dilokpimol and Geshi, 2014).

Our observations thus reinforce the idea that GLCAT14A and GLCAT14C may perform distinct functions, as *glcat14a-1 glcat14c-1* seeds lacking functional GLCAT14A and GLCAT14C exhibited much more severe phenotypes than either of the two single mutant seeds. Notably, a reduction of adherent mucilage was also observed in the CRISPR-Cas9 generated *glcat14abc* mutant (Zhang et al., 2020) as revealed by staining with RR.

Given the striking mucilage phenotype of *glcat14a-1glcat14c-1* seeds in water, we tested for potential mucilage extrusion defects by hydrating seeds in 50mM EDTA pH 8.0 and stainning in RR dye, given that cation chelators like EDTA can disrupt cross-linked unesterified HG chains and facilitate mucilage extrusion (Arsovski et al., 2010; Rautengarten et al., 2008). Interestingly, the mucilage capsules of both *glcat14a-1* and *glcat14c-1* were similar to WT (**Appendix B: Supplemental Figure 3-4A-4C, 4E-4G**); but surprisingly, the *glcat14a-1 glcat14c-1* double mutant seeds lacked any detectable adherent mucilage (**Appendix B: Supplemental Figure 3-4D, 4H**), suggesting that the mucilage phenotype characteristic of *glcat14a-1 glcat14c-1* seeds was due to mucilage repartitioning and not to mucilage extrusion defects.

In addition to the calcium crosslinking of HG to form the "egg-box" structure, Ca^{2+}-driven cross-linking among carboxyl groups of the uronic acid residues within the AGPs and the pectic acids have been speculated (Huang et al., 2016; Lamport et al., 2014). With increased Ca^{2+} ion concentration contributing to mucilage adherence in the adherent layer (Voiniciuc etal., 2013), we analyzed RR stainings of *glcat14* mutant seeds treated with $CaCl_2$. Despite the inherent loss of adherent mucilage in $CaCl_2$ treated *glcat14a-1glcat14c-1* mutant seeds (**Appendix B: Supplemental Figure 3-4L, 4P**), we cannot rule out the possibility that the loss of adherent mucilage mediated by the loss of GlcA residues might have been further exacerbated by a reduction of calcium (Lopez-Hernandez et al., 2020) and thus may contribute to the mucilage defect observed in *glcat14a-1glcat14c-1* mutant seeds.

Both GLCAT14A and GLCAT14C are Required for Mucilage Matrix Polymer Organization and Assembly

Several studies directed at understanding the molecular mechanisms involved in mucilage formation have identified some key players involved in mucilage polysaccharide formation (Griffiths and North, 2017). SOS5/FLA4, the only AGP extensively characterized to date to be involved in mucilage formation, has been implicated in maintaining cell wall structure (Zu et al., 2008; Shi et al., 2003; Harpaz-Saad et al., 2012), and required for mucilage adherence and formation of ray structure (Griffiths et al., 2014). As GlcA is the only acidic sugar in the type II AG glycan with a reported role in calcium binding (Lamport et al., 2014), we demonstrated that GlcA is essential for pectin and cellulose matrix organization in Arabidopsis seed mucilage as revealed by immunolabelling and biochemical analyses. By labelling with calcofluor, a fluorescent probe for cellulose and other β-glucans, we showed the expected combination of intense rays emanating from the top of the columella and diffused staining of rays in the adherent layer for WT, *glcat14a-1* and *glcat14c-1* (**Figure 3-5B**). In addition to the calcofluor labelling around the columella in *glcat14a-1 glcat14c-1* seeds, the diffuse staining of the rays was completely absent. Similarly, in contrast to WT, *glcat14a-1 glcat14c-1* seeds stained with pontamine S4B displayed an irregular distribution of cellulose rays, reduced diffused ray staining between the rays and the incomplete detachment of the outer cell wall in EDTA and water imbibed seeds following pontamine S4B staining (**Figure 3-5A**). Both *sos5* and *fei2* mutant seeds have some mucilage defects such as absence of cellulosic rays, but with intact diffuse staining (Griffiths et al., 2016; Griffiths et al., 2014) that are similar to those observed in *glcat14a-1 glcat14c-1* seeds.

Previous genetic mutant analysis indicated that GALT2, GALT5, SOS5, FEI1, and FEI2 act in a linear, non-additive pathway and suggested that glycosylated SOS5 interacts with FEI1/FEI2 (Basu et al., 2016). Calcium binding of glucuronidated AG polysaccharides in AGPs such as SOS5 may facilitate receptor-ligand interactions necessary for the activity of receptor-like kinases (Lopez-Hernandez et al., 2020). We speculate that the disruption of the *GLCAT14A* and *GLCAT14C* genes may interfere with such interactions and hence associated receptor-like kinase activity. Perhaps there are some AGPs in addition to SOS5/FLA4 that are involved in mucilage formation, and these mucilage AGPs may rely on GLCAT14A and GLCAT14C for biological activity. It is worth mentioning that although 85 AGPs have been identified as members of the superfamily of cell wall proteins (Sullivan et al., 2011), only SOS5 has been implicated in mucilage formation. Also, the reduction in GlcA (**Table 3-1**) may suggest the possible involvement of other GLCATs in mucilage organization and the precise roles of other GLCATs in mucilage formation remains to be elucidated.

Cellulose has been shown to play important roles in anchoring the adherent mucilage to the seed coat (Harpaz-Saad et al., 2012; Sullivan et al., 2011). Aligned cellulose microfibrils in crystalline cellulose produce birefringence of polarized light, and mutants with defects in crystalline cellulose content have been identified based on such altered birefringence (Potikha and Delmer, 1995). The outer epidermal cells of WT and single mutant seeds exhibited strong birefringence with visible rays of crystalline cellulose within the adherent mucilage. By contrast, *glcat14a-1 glcat14c-1* seeds displayed much less birefringence under polarized light, as evidenced by the bright spots on the edges of the seeds (**Figure 3-5E**). This indicates that the crystalline cellulose content was reduced in *glcat14a-1 glcat14c-1* mucilage and thus may have contributed to the loss of adherent mucilage. Understandably, crystalline cellulose was reduced in *cesa5* (Sullivan et al., 2011) and a similar observation was reported for *sos5* (Harpaz-Saad et al., 2012) and *irx14-1* (Hu et al., 2016); but the observed effect of a decrease in β-GlcA content (**Table 3-1**) affecting crystalline cellulose content was rather unexpected. Our findings serve to illustrate how intricately intertwined mucilage polymers are and should be an important consideration in research efforts leading to the deconstruction of plant cell wall assembly processes.

Multiple lines of evidence have revealed potential interactions between Type II AGs and pectin. For example, treatment of cell wall fractions with pectin-degrading enzymes allows for the increased release of AGPs (Immerzeel et al., 2006; Lamport et al., 2006). Similarly, AGPs have been shown to bind to pectins in a calcium-dependent manner (Baldwin et al., 1993). Since glucuronic acid binds to calcium (Lamport et al., 2014), and given the importance of calcium in HG crosslinking and esterification (Voiniciuc et al., 2013), the significant alteration in epitope binding of JIM5 (**Figure 3-**

6B and Appendix B: Supplemental Figure 3-5A2-L2) and JIM7 (**Figure 3-6C;**

Supplemental Figure 3-5A3-L3), especially in *glcat14a-1 glcat14c-1* mutants indicates

that the loss of function of ATGLCAT14A and ATGLCAT14C results in drastic changes

in the HG esterification process. While we observed a reduction in calcium content,

especially in the *glcat14a-1 glcat14c-1* mutants, we only observed an increase in DM in

glcat14a-1 and *glcat14c-1* mutants in total mucilage extracts (**Figure 3-6F**). Surprisingly,

that was not the case for the *glcat14a-1 glcat14c-1* mutants, as a significant decrease in

calcium did not result in increase in mucilage pectin DM. However, we observed an

intense staining of JIM5 and JIM7 tightly bound to the seed coat surface around the

columella for *glcat14a-1 glcat14c-1* mutants (**Figure 3-6B and 6C**). Unfortunately, we

were unable to detect LM2, MAC207 and JIM8 epitopes during whole seed

immunolabelling and immunoblotting experiments; however, JIM13 epitopes were

detected and localized to the columella (**Appendix B: Supplemental Figure 3-5A4-L4**).

JIM13 is a monoclonal antibody that detects an AGP-related glycan, specifically the

epitope: β-D-GlcA-(1,3)-α-D-GalA-(1,2)-α-L-Rha; (Knox et al., 1991). This finding lends

credence to an earlier observation that showed the presence of AGPs in mucilage

(Griffiths et al., 2014). Given the reported role of glucuronic acid of AGPs in stabilizing

the covalent attachment of rhamnosyl residues of pectin RG-I backbone to AGPs in

APAP1 (Tan et al., 2013), the reduction or loss of GlcA residues in AGPs may have

contributed to the loosely held mucilage of the adherent layer in *glcat14a-1 glcat14c-1*

mutants being released upon gentle shaking in water (**Figure 3-3C**).

Role of GLCAT14A and GLCAT14C in Seed Coat Epidermal Cell Development

The surface morphology of mature WT and *glcat14a-1 glcat14c-1* seeds were examined by SEM to investigate whether β-GlcA has a role in seed coat epidermal (SCE) cell development. We observed that SCE cells in *glcat14a-1 glcat14c-1* seeds were deformed with the collapse of their polygonal structures (**Figure 3-8A and 8B**). Specifically, we observed a collapse of the radial cell wall coupled with an increase in the size of the columella (**Figure 3-8C-8F**). Western *et al.* (2004) suggested that a decrease in the amount of mucilage synthesized results from a smaller mucilage pocket and a much flatter columella. That appears to be the case here, as our data showed an increase in columella size before mucilage extrusion and remains unchanged following mucilage extrusion. This might explain the reason for the reduction in total mucilage content for the *glcat14a-1 glcat14c-1* mutant (**Figure 3-7**) advanced by smaller mucilage pockets that are known to precede columella formation (Griffiths et al., 2014). Evidently, CESA2, CESA5 and CESA9 are involved in radial cell wall reinforcement and columella deposition (Stork et al., 2010; Mendu et al., 2011), but it remains to be determined whether the reduction in crystalline cellulose content in *glcat14a-1 glcat14c-1* seeds might have impacted the columella formation. Notably, fully glycosylated FLA4/SOS5 molecules were identified to be candidates transported to the plasma membrane while insufficiently *O*-glycosylated protein regions are targeted for vacuolar degradation (Seifert, 2020). In that case, the modification of SOS5 glycans (i.e., the loss of glucuronic acid) might have interfered with SOS5's intracellular trafficking (Seifert, 2020) and its proposed interactions with the FEI ectopic domain thus affecting mucilage polymer assembly and SCE cell formation (**Figure 3-10**).

Figure 3-10

Proposed Model for the Potential Role of ATGLCAT14A and ATGLCAT14C in Arabidopsis Mucilage Secretory Cells During Mucilage Formation

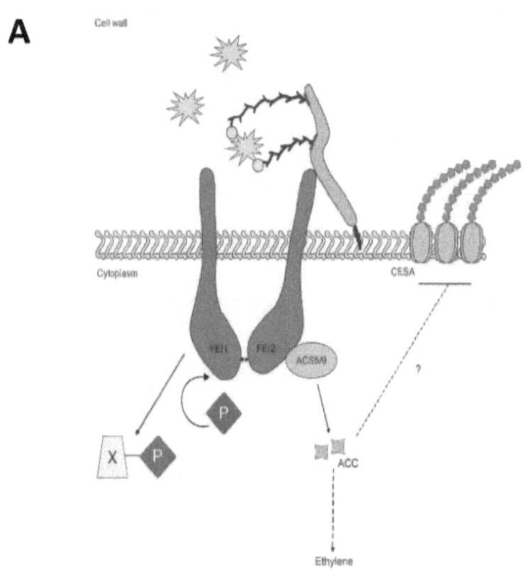

Figure 3-10: continued

B

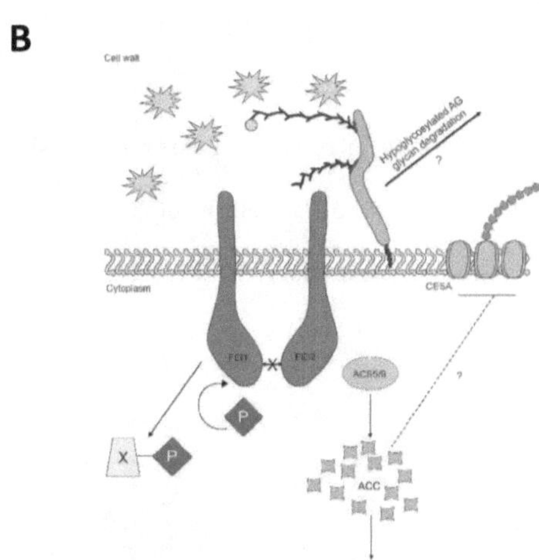

Note. In A, the interaction of glucuronidated AG glycan and Ca^{2+} may stabilize the interaction between AGPs (e.g., SOS5) and FEI receptor kinases leading to the activation of the FEI kinase domain. FEI1/FE2 binding to ACC synthase (ACS5/9) limits the production of ACC, and influences either directly or indirectly, the formation of cellulose microfibril assembly, independent of ethylene. In B, the absence or reduction of glucuronidated AG glycans may interfere with and reduce Ca^{2+} binding and may destabilize the activities of FEI receptor like proteins by altering potential interactions between AGPs and FEI. ACC synthase remains unbound to FEI proteins and becomes freely available for the increased production of ACC, which inhibits cellulose synthesis and increases the production of ethylene. Similarly, hypoglycosylated moieties characterized by AG glycan modifications are targeted for destruction to the vacuoles via multivesicular bodies.

We observed a significant reduction in seed size for *glcat14a-1glcat14c-1* mutants
(**Figure 3-9A-9C**) relative to WT. Although an increase in Gal was observed for
*glcat14a-1 glcat14c-1*seeds relative to WT, this increase was found not to be significant.
Therefore, further analysis is needed to validate the plausible hypothesis that β-GlcA
terminates the elongation of β-(1→6)-galactan side chains (Knoch et al., 2013).

Chapter 4: Three β-Glucuronosyltransferase Genes Involved in Arabinogalactan Biosynthesis Function in Arabidopsis Growth and Development

The work in this chapter was published and corresponds to the following journal citation:

Ajayi, O.O., Held, M. A., and Showalter, A.M (2021) Three β-Glucuronosyltransferase Genes Involved in Arabinogalactan Biosynthesis Function in Arabidopsis Growth and Development. *Plants*, 10(6): 1172

Introduction

Plant cell walls are comprised of structurally complex heteropolymers that are essential to plant development. Arabinogalactan-proteins (AGPs) are a family of complex proteoglycans found in the plant cell walls of all higher plants (Gaspar et al., 2001) and contain a large amount of arabinogalactan polysaccharide (90% w/w) attached to hydroxyproline residues in their protein cores (Showalter et al., 2010). AGPs belong to a superfamily of structural macromolecules known as the hydroxyproline -rich glycoproteins (HRGPs) to which EXTs and proline-rich proteins are members with unique carbohydrate moieties central to their biological functions (Kieliszewski, 2001). Out of all the HRGPs, AGPs are the most complex, and their complexity is driven by the heterogeneity of their type II arabinogalactans (AG) and the variety of their protein backbones (Ellis et al., 2010). AGPs have N-terminal signal peptides, which are cleaved to allow for secretion, and about half of AGP family members also have C-terminal glycosylphosphatidylinositol (GPI) anchors to allow for their retention and attachment to the outer leaflet of the plasma membrane (Seifert and Roberts, 2007).

AGPs are implicated in an array of plant growth and development processes including cell expansion, somatic embryogenesis, root and stem growth, salt tolerance, hormone signaling, programmed cell death, male and female gametophyte development, and wounding/defense (Ellis et al., 2010; Showalter, 2001). Structurally, AGs are O – linked to hydroxyproline residues in the AGP protein core and each AG is characterized by the presence of a β-1,3-galactan backbone that is substituted at various $O6$ positions with β-1,6-galactan side chains (Knoch et al., 2013). These so-called type II galactan chains are further modified by arabinose (Ara), rhamnose (Rha), fucose (Fuc), xylose (Xyl) and glucuronic acid or 4–O–methyl glucuronic acid (collectively referred to as GlcA) (Tan et al., 2010; Tryfona et al., 2010; Tryfona et al., 2012). These various AGP glycosylation reactions are catalyzed by a family of enzymes known as glycosyltransferases (GTs) (Tryfona et al., 2010). While the functional roles of several AGP GTs have been identified and characterized to varying extents using molecular and biochemical approaches, the identity and biological roles of many other AGPs GTs remain to be explored.

The AGP glucuronosyltransferases or glucuronic acid transferases (GLCATs) represent one family of AGP GTs, which has been characterized to some extent. As their name implies, this family of enzymes transfer glucuronic acid (GlcA) to type II AGs and are members of the GT14 family of the Carbohydrate Active Enzyme (CAZy). Previous work on GLCAT14A-C demonstrated that they transfer GlcA to both β-1,3-galactan and β-1,6-galactan side chains with distinct substrate preferences (Knoch et al., 2013; Dilokpimol and Geshi, 2014). Recently, two additional genes/enzymes, GLCAT14D and E were reported to be involved in the glucuronidation of type II AGs (Lopez-Hernandez et al., 2020). Furthermore, *glcat14a-1* and *glcat14a-2*, two allelic T-DNA insertion mutants, were reported to display increased cell elongation phenotype in etiolated seedlings (Knoch et al., 2013), a finding that was disputed in a later study (Lopez-Hernandez et al., 2020). In addition, multiple developmental defects such as reduced trichome branching and severe reduction in growth of etiolated seedlings were observed in *glcat14a/b/e* mutants, including the perturbation of the calcium waves (Lopez-Hernandez et al., 2020). In another related study, CRISPR-Cas9 generated *glcat14a/b* and *glcat14a/b/c* mutants showed delayed seed germination, reductions in root hair length and trichome branching, reduced seed set and developmental defects in seed mucilage and pollen grains (Zhang et al., 2020). These results reinforce the critical roles that these biosynthetic enzymes play in different aspects of plant growth.

Calcium is an essential element needed for growth and development of plants under both normal (non-stress) and stress conditions. Calcium impacts cell wall rigidity, and low calcium content weakens the cell wall to facilitate cell expansion (Bascom et al., 2018). Notably, calcium performs multiple roles in plant growth, but its effect is not limited to cell wall and membrane stability, but also serves as a secondary messenger in many aspects of plant biological processes (Thor, 2019). There are speculations that AG glucuronidation of AGPs is a major source of cytosolic Ca^{2+}, given the negatively charged GlcAs residues' inherent ability to bind and release calcium in a pH-dependent manner at the periplasmic surface of the plasma membrane (Lamport and Varnai, 2013). Previous work has showed that some of the developmental defects in *glcat14a/b/e* mutants, specifically the trichome branching defects, were suppressed by increasing the calcium content in the growth medium (Lopez-Hernandez et al., 2020). This reinforces the interdependence between cell surface apoplastic calcium and GlcA residues in AGPs in regulating important signaling events critical to plant growth and development.

Here, we focus on the biochemical and physiological characterization of *glcat14* T-DNA insertion mutants of *glcat14a*, *glcat14b*, *glcat14c*, *glcat14a/c*, *glcat14b/c* along with CRISPR-Cas9 generated *glcat14a/b* and *glcat14a/b/c* mutants. Specifically, we sought to understand the degree to which the alterations in the amounts of GlcA and calcium content accounts for the mutant phenotypes identified in this study using various biochemical and molecular genetic approaches. This will bring us a step closer towards understanding the relative contributions of the *GLCAT* genes in cell wall integrity maintenance critical to plant growth processes.

Materials and Methods

Plant Materials

Arabidopsis T–DNA insertion lines *atglcat14a–1* (SALK_064313) and *atglcat14a–2* (SALK_043905), *atglcat14b-1* (SALK_080923) and *atglcat14b-2* (SALK_117005), *atglcat14c-1* (SALK_005705) and *atglcat14c-2* (SALK_051810) were selected using the SIGnaL database (http://signal.salk.edu/) and were obtained from the ABRC (Arabidopsis Biological Research Centre) (http://abrc.osu.edu/) (Alonso et al., 2003). We could not verify the presence of a T-DNA insertion in SALK_051810, but a CRISPR knock-out mutant (*glcat14c-2*) in the first exon generated a 178 bp deletion (Zhang et al., 2020) and was included in this study as an allelic mutant that exhibited similar phenotypes to *atglcat14c-1* (SALK_005705). Arabidopsis ecotype Col–0 was used as the WT for comparison. Homozygous lines were identified by PCR using primer sets identified using the T-DNA primer design tool provided by the Salk Institute (**Figure 4-1A; Appendix A: Supplemental Table 4-1**).

Homozygous *glcat14a–1* and *glcat14c-1* were crossed, and the F1 generation was selfed before screening the F2 generation to obtain *glcat14a/c*. Similarly, *glcat14b–1* and *glcat14c-1* were crossed, and the F1 generation was selfed before screening the F2 generation to obtain *glcat14b/c* (**Appendix B: Supplemental Figure 4-1**). CRISPR-Cas9 generated *glcat14a/b* and *glcat14a/b/c* mutants (henceforth, referred simply as *glcat14a/b* and *glcat14a/b/c*) were previously generated in our lab (Zhang et al., 2020) and included in this study to gain broader insight into the biological roles of the GLCATs. For the purpose of this work, *glcat14a/b* and *glcat14a/b/c*, generated by CRISPR-Cas9 approach are referred to as *glcat14* CRISPR lines while the other genotypes involving T-DNA insertion mutants are referred to as *glcat14* T-DNA insertion lines.

Quantitative RT-PCR

To verify whether gene transcripts exist in Arabidopsis *glcat14* T-DNA insertion lines, total RNA was extracted from Arabidopsis leaves using Trizol (Life Technologies, Grand Island, NY, USA). RNA (1 μg) was used for first-strand cDNA synthesis along with an oligo (dT20) primer and SuperScript III reverse transcriptase (Thermo Scientific). The qPCR was performed using PerfeCTa SYBR Green SuperMix (Quanta Biosciences) following methods described elsewhere (Zhang et al., 2020). Expression levels were calculated relative to the Arabidopsis *ACTIN2* gene.

Figure 4-1

GLCAT14A, GLCAT14B, and GLCAT14C Gene/Genetic Mutant Structure, Phylogenetic

Relationships, and Expression

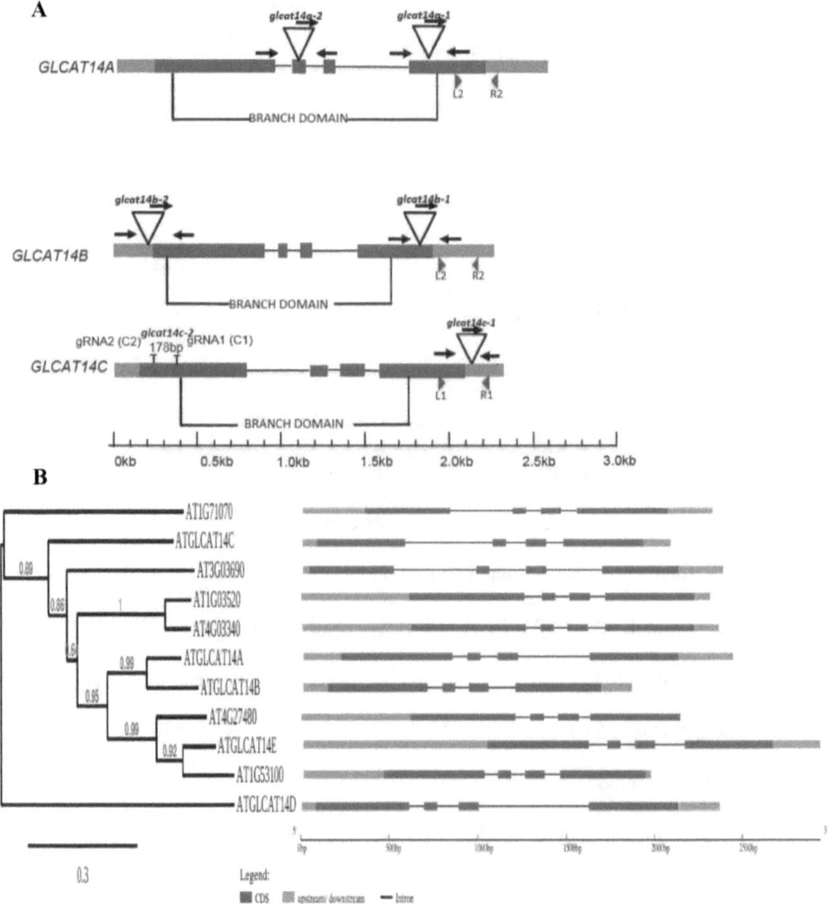

Figure 4-1: continued

C

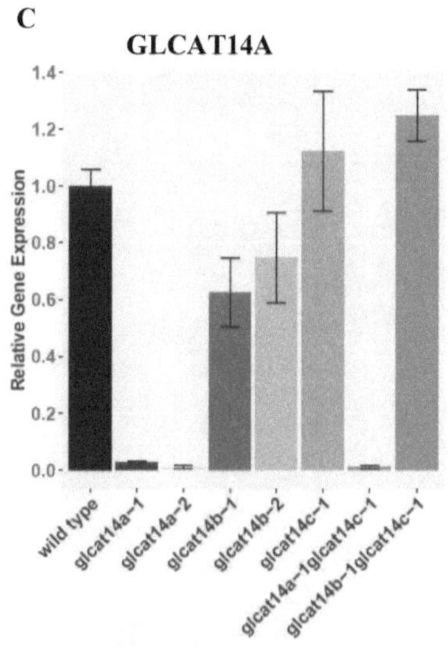

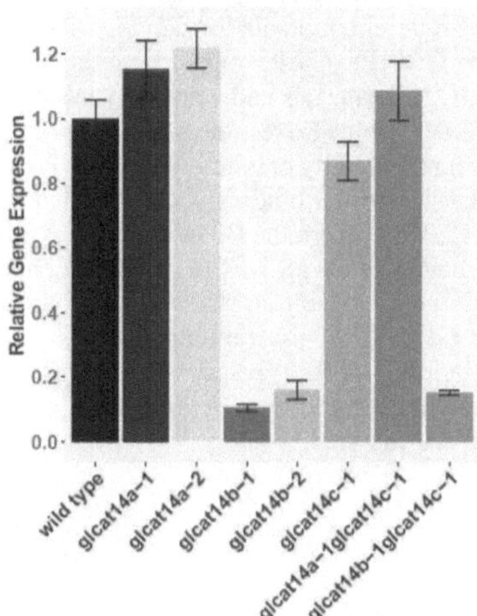

Figure 4-1: continued

GLCAT14C

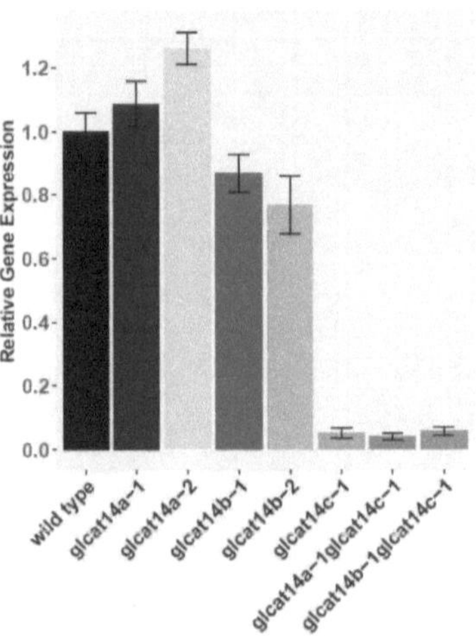

Note. A, Schematic drawing of the GLCAT14A/B/C gene structure and sites of T-DNA insertions in two independent lines: *atglcat14a-1* and *atglcat14a-2* are allelic mutants for *GLCAT14A*; *atglcat14b-1* and *atglcat14b-2* are allelic mutants for *GLCAT14B*; *atglcat14c-1*, and *atglcat14c-2*, generated by CRISPR-Cas9 strategy targeting exon 1 by two guide RNAs gRNA1 (C1) and gRNA2 (C2) (Materials and Methods) are allelic mutants for *GLCAT14C*. Exons are represented by blue boxes; introns are represented by thin gray lines and non-coding regions are represented by orange boxes. The T-DNA insertion sites for each mutant are indicated, with primer binding sites indicated by black arrows. The branch domain corresponds to PF02485 from the Pfam database is indicated above. Red arrowheads indicate primer binding sites for qRT-PCR. B, Phylogenetic analysis of the GT14 gene family along with their exon-intron structure. C, Relative gene expression of *GLCAT14A*, *GLCAT14B*, and *GLCAT14C* in Arabidopsis leaves in various genetic mutant backgrounds as determined by qRT-PCR. Transcript levels were normalized to the mean of one reference gene, the Arabidopsis *Actin 2* gene, *AtACT2*. Averages of three biological replicates ± SE are shown.

Bioinformatics

Phylogenetic analysis of CAZy GT14 family members was conducted using PhyML (http://www.phylogeny.fr/one_task.cgi?task_type=phyml) while the Gene Structure Display Server was used to determine the genetic architecture of Arabidopsis GT14 gene family (http://gsds.gao-lab.org/).

AGPs Quantification Using β-D-Gal-Yariv Reagent

AGPs were extracted from rosette leaves, stems, and siliques of 40-day-old WT, *glcat14* T-DNA insertion and CRISPR lines, using the β-D-Gal-Yariv precipitation method described elsewhere (Lamport, 2013). The dissolved AGPs were quantified by measuring absorbance at OD_{420}. Different concentrations of gum arabic (G9752, Sigma-Aldrich, St. Louis, MO, USA) dissolved in 1% $CaCl_2$ were used to make the standard curve. Measurements for each genotype were done in triplicate.

Monosaccharide Composition Analysis by High Performance Anion Exchange Chromatography with Pulsed Amperometric Detection (HPAE-PAD)

Following the extraction of AGPs as described above, fifty microliters of 10 mg/mL AGP were hydrolyzed using 2 N TFA at 121°C for 90 min. TFA was removed by evaporation with N_2 gas. Samples were dissolved in 500 μL milli-Q water containing 0.2 mM cellobiose as an internal standard. A standard sugar mixture (fucose, rhamnose, arabinose, galactose, glucose, xylose, mannose, galacturonic acid, and glucuronic acid) were used for making the standard curve.

Monosaccharide compositions were calculated as molar percentages (mol %). All samples and standards were subjected to high pH anion-exchange chromatography with pulsed amperometric detection (HPAE-PAD) using a Dionex PA-20 column (Thermo Fisher Scientific, Sunnyvale, CA, USA) essentially as described previously (ØBro et al., 2004).

Calcium Binding Assay

Following the extraction of AGPs, calcium binding assays were conducted as previously described (Zhang et al., 2020). Briefly, 10 µL of 10mg/ml AGP extracted from *glcat14* T-DNA insertion and *glcat14* CRISPR lines were used for the calcium binding assay. A commercial calcium colorimetric assay kit (MAK022, Sigma-Aldrich, St. Louis, MO, USA) was used for calcium measurement following the manufacturer's protocol. In this assay, calcium ions from the AGP extracts form a complex with the o-cresolphthalein, which causes a color change from transparent to pink. The amount of calcium was determined using a UV spectrometer at OD_{575} and a standard curve made with different concentrations of $CaCl_2$.

Seed Germination

One-month post-harvested seeds of WT and allelic *glcat14* mutant lines (*glcat14a-1, glcat14a-2, glcat14b-1, glcat14b-2, glcat14c-1, glcat14c-2, glcat14a/c, glcat14b/c, glcat14a/b,* and *glcat14a/b/c*) were used for germination experiments. Following seed sterilization, seeds were stratified at 4°C for 3 d in the dark and were sown on ½ MS and 1% sucrose agar plates supplemented with or without 1 µM ABA. Germination percentages were counted from 1 to 12 d after sowing and approximately 25 seeds were sown for each genotype with three replicates. In another experiment to evaluate the growth of etiolated seedlings of respective genotypes, stratified seeds sown on MS plates were exposed to 4 h of fluorescent white light at 20°C to synchronize germination before wrapping the plates in aluminum foil for growth in the darkness for 5 d at 20°C. Root and hypocotyl length measurements at 5 d were taken and values were compared to WT.

Trichome Morphology

The first two leaves of 14-day-old WT, *glcat14a/b* and *glcat14a/b/c* mutants were immersed in methanol and leaf tissues were critical point dried following methods described earlier (Talbot and White, 2013) before examining using scanning electron micrography (SEM). Trichomes were examined using a JEOL JSM-6390 scanning electron microscope (Hitachi High-Technologies) and leaf tissues were mounted on aluminum stubs using double adhesive tapestubs and sputter coated with a palladium alloy using a Cressington 208C high-resolution sputter coater (Ted Pella Inc.) at the Institute for Corrosion and Multiphase Technology, Ohio University. The Scanning Electron Microscope Energy Dispersive X-ray (SEM/EDX) was used as a semi quantitative approach to estimate the calcium content in the trichomes of WT, *glcat14a/b* and *glcat14a/b/c* mutants following methods described previously (Scimeca et al., 2018).

In vitro Pollen Germination

Flowers from 30-day-old WT and *glcat14a*, *glcat14b*, *glcat14c*, *glcat14a/c*, *glcat14b/c* plants were used for an *in vitro* pollen germination assay. Pollen germination medium contained 10% sucrose, 0.01% boric acid, 1 mM $CaCl_2$, 1 mM $Ca(NO3)_2$, 1 mM KCl, 0.03% casein enzymatic hydrolysate, 0.01% myo-inositol, 0.1 mM spermidine, 10 mM GABA, 500 μM methyl jasmonate, and 1% low-melting agarose. Pollen from each genotype was incubated on pollen germination medium. Pollen shape, pollen germination rate, and pollen tube length were measured 3 h after incubation with a Nikon phot-lab2 microscope at 10x magnification. More than 150 pollen and pollen tubes were measured in an individual experiment with three replicates.

Western Blotting Analysis

The total protein extraction from leaves and stems of WT, *glcat14a/b* and *glcat14a/b/c* mutants were performed following methods described previously (Fragkostefanakis et al., 2012). Total protein supernatants from the samples extracted were quantified by the Bradford method (Bradford, 1976). Twenty-five micrograms (25µg) of crude protein extract of WT, *glcat14a/b* and *glcat14a/b/c* mutants were separated by 10% mini-PROTEAN® TGX™ (456-1033; Bio-Rad, **https://www.bio-rad.com**) gel electrophoresis and transferred onto a 0.45-µm Immun-Blot low florescence PVDF membrane (1620260; Bio-Rad) using the wet transfer system at a constant current of 100V for 45 min. The blots were blocked for 1 h (3% BSA, in 1× Tris-Buffered Saline with Tween [150 mM NaCl, 20 mM Tris base, 0.1% Tween 20; TBST]), probed with 1:10 LM2 monoclonal rat IgM primary antibody (Carbosource, **https://www.ccrc.uga.edu**) in blocking buffer for 1 hr, washed three times with 1× TBST for 15 min, probed with goat anti-rat IgG H + L secondary antibody (PI31629; FisherScientific), washed thrice with 1× TBST for 10 min, once with 1× TBS for 10 min, and treated with chemiluminescent substrate (Clarity™ ECL substrate, 1705060S; Biorad). A ChemiDoc imaging system (Bio-Rad) was used for image acquisition. PVDF membranes were incubated for 10 min with Ponceau S solution [0.1% (w/v) Ponceau S; 5% (v/v) acetic acid] to stain for total proteins.

Cloning Procedures and Subcellular Localizations of GLCAT14A, GLCAT14B, and GLCAT14C

Full length *sequenced ATGLCAT14A, ATGLCAT14B and ATGLCAT14C CDS*, flanked by the sites *KpnI* were subcloned into pSAT6 vector to generate pSAT6-*GLCAT14A*-EYFP and pSAT6-*GLCAT14C*-EYFP constructs containing *PI-PSPI* restriction sites. Following the digestion of the new constructs (pSAT6-*GLCAT14A*-EYFP, pSAT6-*GLCAT14B*-EYFP and pSAT6-*GLCAT14C*-EYFP) with *PI-PSPI*, the fragment was introduced into the pPZP-RCS-nptII binary vector (Citovsky et al., 2006). The binary vector containing the constructs were transformed into *Agrobacterium* strain GV3101 before infiltration using 5-week old tobacco leaves grown at 22–24 °C (Batoko et al., 2000). Golgi co-localisation was confirmed by using the Golgi marker sialyltransferase short cytoplasmic tail and single transmembrane domain fused to enhanced GFP (STtmd-GFP). Transformed plants were incubated under normal growth conditions and imaged 3 days post- infiltration using an upright Zeiss LSM 510 META laser scanning confocal microscope (Jena, Germany). Both GFP and E-YFP were imaged using 458-nm and 488-nm excitation wavelength respectively

Results

The CAZy GT14 Family

This work sought to determine the biochemical and physiological roles of three functionally characterized GLCATs (GLCAT14A, GLCAT14B and GLCAT14C) in

Arabidopsis. Previous work provided enzymatic evidence that three GLCATs (GLCAT14A, GLCAT14B, GLCAT14C) out of the 11 putative Arabidopsis GLCAT genes/enzymes in the CAZy GT14 family function in glucuronidation of AGPs (Dilokpimol and Geshi, 2014). More recently, GLCAT14D and GLCAT14E were reported to be involved in AG glucuronidation (Lopez-Hernandez et al., 2020). A common feature in verified and putative GLCATs in Arabidopsis is the presence of a highly conserved GLCAT domain (Ajayi and Showalter, 2020). Also, the intron-exon genetic architecture of Arabidopsis CAZy GT14 family members is conserved and includes the presence of 4 exons and 3 introns (**Figure 4-1B**). Phylogenetic analysis showed that *GLCAT14A* and *GLCAT14B* belong to the same phylogenetic clade while *GLCAT14C, D* and *E* are phylogenetically distant from *GLCAT14A* and *GLCAT14B*, consistent with earlier reports (Dilokpimol and Geshi, 2014 and Lopez-Hernandez et al., 2020). Sequence analysis showed differences in similarities among the *GLCAT14* genes with the lowest sequence similarity observed to be 39% (between *AT1G53100* and *ATGLCAT14D*) and highest (between *AT4G03340* and *AT1G03520*); others are nestled between these two extremes (**Appendix A: Supplemental Table 4-2**). Interestingly, these five enzymes, GLCAT14A/B/C/D/E, demonstrated to be involved in the glucuronidation of type II AGs, have unique substrate preferences; GLCAT14A, GLCAT14B and GLCAT14C transfer GlcA onto both β- 1,3 and β- 1,6 galactan chains of shorter length, while GLCAT14D and GLCAT14E have preferences for longer galactan side chain (Dilokpimol and Geshi, 2014; Lopez-Hernandez et al., 2020).

glcat14 Mutant Generation and Characterization

Arabidopsis *glcat14* T-DNA insertion mutant lines were confirmed for the presence of T-DNA insertions (**Appendix B: Supplemental Figure 4-1A**) and were subsequently used for genetic crosses to obtain higher order homozygous mutants (**Appendix B: Supplemental Figure 4-1B and 1C**). Also, results obtained using qRT-PCR showed the absence of transcripts in these mutants (**Figure 4-1C**). We could not verify the presence of a T-DNA insertion in the SALK_051810 line belonging to *GLCAT14C*, but a CRISPR knock-out mutant (*glcat14c-2*) has a 178 bp deletion (Zhang et al., 2020) in the first exon (**Figure 4-1A**) and biological phenotypes were comparable to those obtained in the *glcat14c-1* mutant (SALK_005705) as subsequently described.

glcat14a/b and glcat14a/b/c Had Reduced Immunolabelling of LM2-bound AGPs in Stem and Leaf

Structural modifications can affect the abundance of glycan moieties that make up the plant cell wall. These structural changes can be delineated using antibodies raised against glycan epitopes in an immunoblot analysis. The LM2 antibody, which specifically binds to β-glucuronic acid in Type II AGs in AGP protein cores, was used to immunolabel WT, *glcat14a/b* and *glcat14a/b/c* leaf and stem protein extracts separated by SDS-PAGE prior to membrane transfer. As expected, LM2 immunolabelling generated broad smears, reflective of the heavily glycosylated nature of AGPs, and the labelling intensities of LM2 bound AGPs were greater in leaf than in stem (**Figure 4-2A and 2B**). In contrast to the WT, weak signals with LM2 bound AGPs were observed in *glcat14a/b* and *glcat14a/b/c* extracts from leaf (**Figure 4-2A**) and stem (**Figure 4-2B**).

Moreover, the molecular size of LM2 bound AGPs in the stem and the leaf were similar, and signal intensities were in the high molecular mass (100 - 250 kDa) region of the western blot.

Figure 4-2

Western Blot Analysis of Protein Extracts from Leaf (A) and Stem (B) of WT, glcat14a/b and glcat14a/b/c

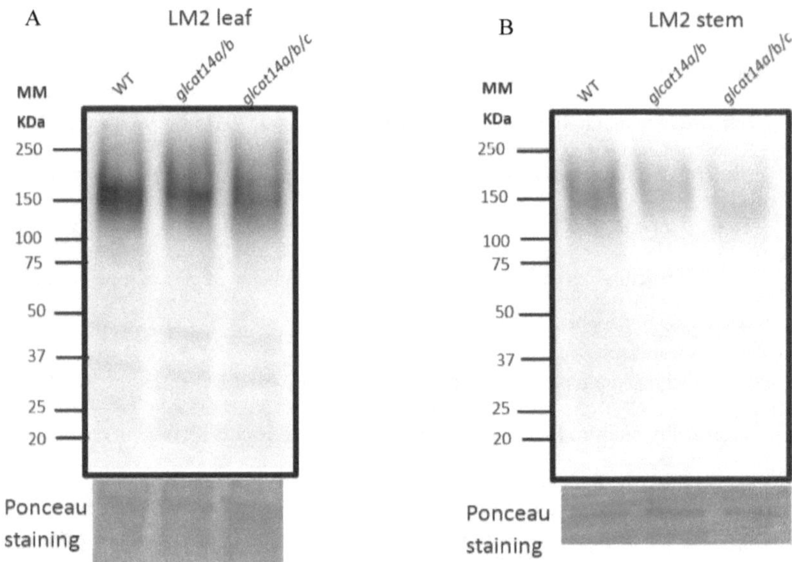

Note. Protein (25μg) extracted from plant materials were loaded per lane and separated by SDS-PAGE according to size, ponceau stained, and subsequently transferred to PVDF membranes. LM2 antibody was used for immunoblotting to target β-GlcA in leaf (A) and stem (B). Molecular mass (kDa) is indicated on the left. MM, Molecular marker.

Quantification of β-D-Gal-Yariv Precipitated AGPs in glcat14 Mutants

Given the differential expression of *GLCATs* in different organs and across different developmental stages (Ajayi and Showalter, 2020), we quantified the amount of Yariv precipitable AGPs from different organs of WT and *glcat14* mutants. The ability of β-D-Gal-Yariv to bind to the β-1,3-galactose backbone on AGPs was exploited to isolate and quantify AGP content as described previously (Kitazawa et al., 2013). Across the organs examined (rosette leaves, stems and siliques), the amount of Yariv precipitable AGPs in the stems and siliques were greater than those observed in the rosette leaves. Specifically, in contrast to WT, we observed an increase in Yariv precipitable AGPs in *glcat14a, glcat14a/c, glcat14a/b* and *glcat14a/b/c* mutants in the rosette leaves and stems; however, in the siliques, the *glcat14a* mutants along with *glcat14c, glcat14a/c, glcat14a/b* and *glcat14a/b/c* mutants showed an increase in Yariv precipitable AGPs (**Figure 4-3**). The *glcat14b* and *glcat14b/c* mutants, however, did not show an increase in AGP content and were comparable to WT.

Figure 4-3

Quantification of AGPs in the Various glcat14 Mutants in Different Organs of 40-Day-Old Arabidopsis Plants

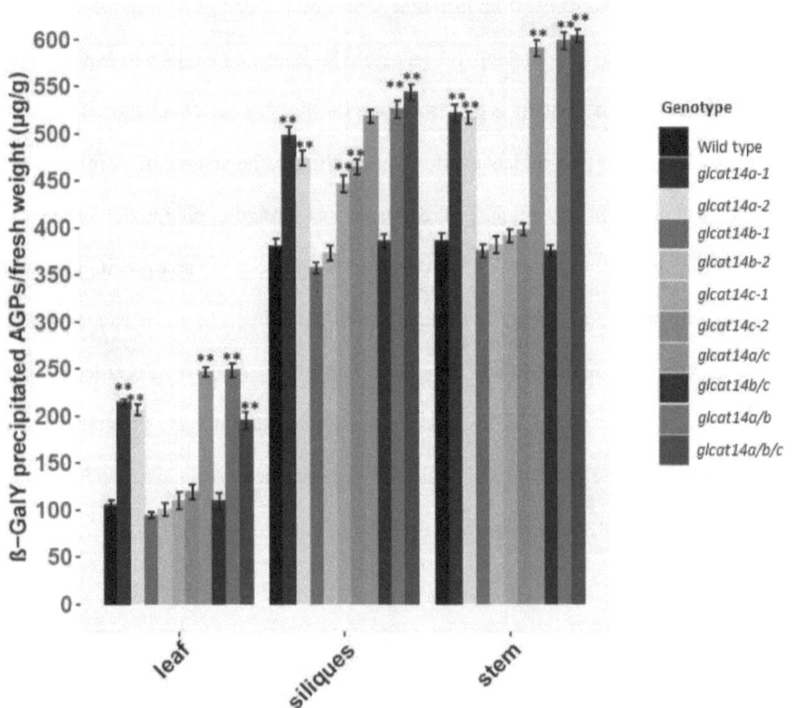

Note. Leaves, siliques, and stems were obtained from 40-day-old Arabidopsis plants. *glcat14a, glcat14a/c, glcat14a/b and glcat14a/b/c* showed significant increases in AGP content in the leaves and stems, while *glcat14a, glcat14c, glcat14a/c, glcat14a/b and glcat14a/b/c* showed increases in AGP content in siliques. AGPs were measured as microgram per gram fresh weight. Statistical differences were determined by two-way ANOVA, followed by the Tukey's honestly significant difference test (** $P < 0.01$); results were averages of three independent experiments.

Monosaccharide Composition Analyses and Calcium Content

AGP glycans were evaluated in stems, siliques and leaves to provide insights into organ specific glycan composition and to understand to what extent the AGP structural modifications would impact the abundance of other AGP sugars in *glcat14* mutants relative to WT. When the plant cell wall integrity (CWI) system is compromised, adaptive changes in cellular and cell wall metabolism can alter wall composition and structure (Gigli-Bisceglia et al., 2020). Previous work on *glcat14a* mutants reported a slight increase in galactosylation while the GlcA content was comparable to WT in AG extracts from 14-day-old seedlings (Knoch et al., 2013). In acid-hydrolyzed AGP extracts investigated by HPAE-PAD, we observed significant changes in sugar amounts relative to WT. Specifically, in the leaves, and with the exception of *glcat14a* and *glcat14c*, we observed an increase in the amounts of galactose and minor changes in arabinose relative to the WT (**Appendix A: Supplemental Table 4-3**). While reductions in the amount of GlcA were observed across all genotypes in leaves, we focused on *glcat14a/b* and *glcat14a/b/c* mutants because the leaf trichome branching phenotype were only found in these mutants (discussed below) and was absent in the *glcat14* T-DNA insertion lines. In the leaf AGPs, we observed a 47% and 49% reduction in the amounts of GlcA in *glcat14a/b* and *glcat14a/b/c* mutants, respectively, relative to WT (**Figure 4-4A**), while for stem AGPs, we observed a 35% and 73% reduction in GlcA content in *glcat14a/b* and *glcat14a/b/c* mutants, respectively, relative to WT; the remaining genotypes had comparable GlcA amounts to WT in stem AGPs (**Appendix A: Supplemental Table 4-4**). With the exception of *glcat14a/c* and *glcat14b/c*, which were not significantly different from WT, a reduction in galactose content in *glcat14a/b/c* in the stem was unexpected, given the significant increases in AGP content of *glcat14a/b/c* mutants in stem tissues. In addition, minor changes were observed in the amount of arabinose in *glcat14* mutants relative to WT. For the siliques, we observed a reduction in the GlcA content in *glcat14* mutants, coupled with a corresponding increase in galactose content in both single and higher order *glcat14* mutants (**Appendix A: Supplemental Table 4-5A**). It should be noted that some residue amount of GlcA remains associated with the AGPs in all of these organs, regardless of the mutant. This suggest the involvement of other GLCATs, whose roles remain to be elucidated. Calcium is well known as a universal signaling molecule and acts as a 'second messenger' in plants (Hepler, 2005; Vanneste and Friml, 2013). The ability of AGP to bind and release calcium in a pH-dependent manner *in vitro* gave rise to the AGP- Ca^{2+} capacitor hypothesis (Lamport and Varnai, 2013). The intramolecular calcium binding abilities by paired GlcA residues of AGPs was exploited to estimate the calcium content of AGPs. We investigated whether there were significant alterations in calcium bound by AGPs, possibly mediated by the reduction in the amounts of GlcA across tissues investigated using a colorimetric assay (Corns and Ludman, 1987). We observed that the calcium content of AGP extracts in *glcat14* mutants were similar to the WT in leaves (**Figure 4-4B**) and siliques (**Appendix A: Supplemental Table 4-5B**); however, that was not the case for stems.

Figure 4-4

Biochemical and Physiological Analyses of WT and glcat14 Mutants

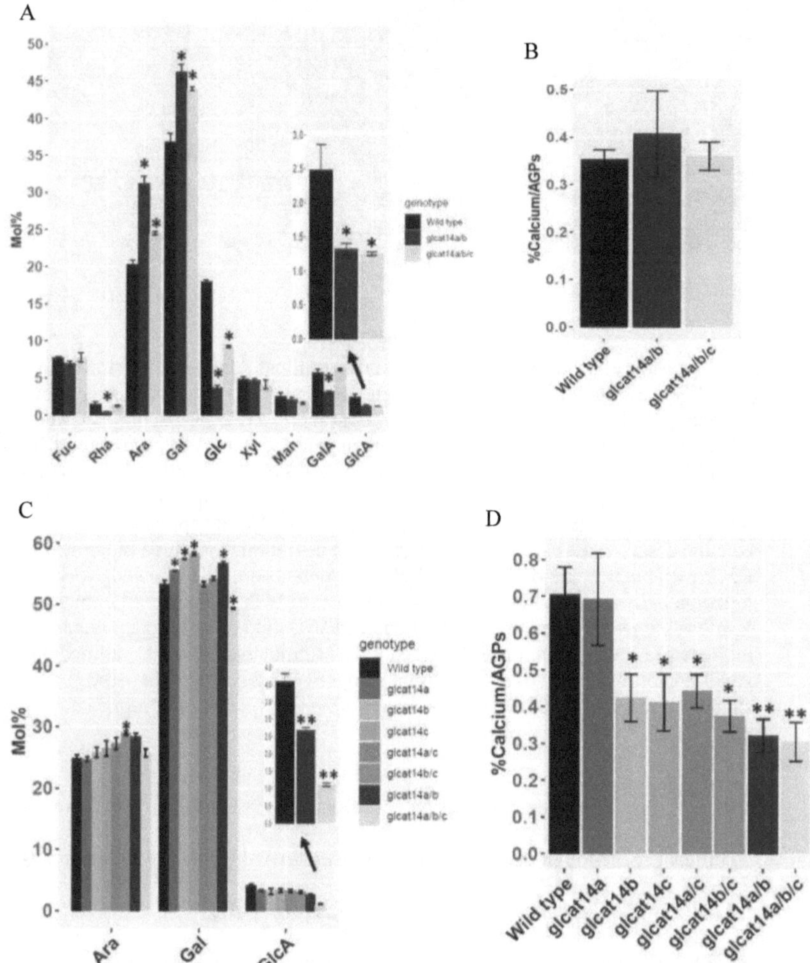

Figure 4-4: continued

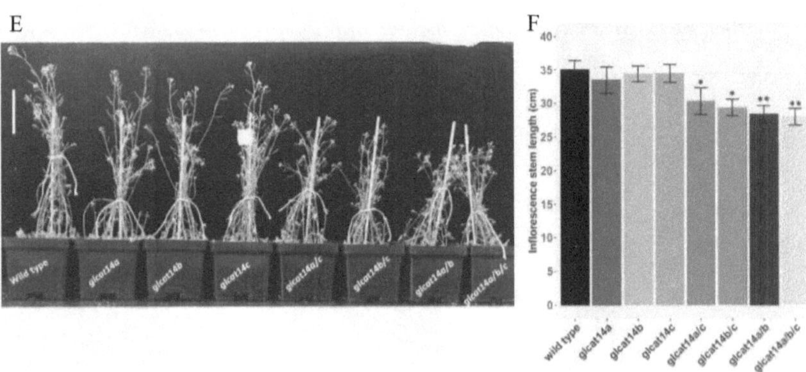

Note. A, Monosaccharide composition analysis of leaf AGPs extracted from WT and *glcat14a/b* and *glcat14a/b/c* mutants using HPAE-PAD. B, Calcium content of leaf AGP extracts of WT, *glcat14a/b*, and *glcat14a/b/c* (expressed as % calcium per AGPs). C, HPAE-PAD monosaccharide composition analysis of stem AGPs extracted from WT and *glcat14* mutants. D, Calcium content of stem AGPs extracted from WT, *glcat14a/b*, and *glcat14a/b/c* (expressed as % calcium per AGPs). For the sugar analysis, values are relative to total sugar composition (expressed as mol %) of triplicate assays ± SE. For the calcium assay, values are the mean ± standard deviation from three biological replicates. E, Growth phenotypes of 40-day-old *glcat14* mutants and WT showed growth reductions in the *glcat14* double and triple mutants relative to WT; Scale bar = 12 cm. F,Inflorescence stem lengths from 40-day-old WT and *glcat14* mutant plants. Asterisks indicate significant differences between *glcat14* mutants and WT as defined by one-way ANOVA followed by Tukey's multiple comparison test (Student's *t*-test, P < 0.05 for single asterisks, P < 0.01 for double asterisks), $n = 15$ per line per replicate.

While the calcium content of *glcat14a* was similar to WT, we observed a significant reduction in calcium content in *glcat14b* and *glcat14c* single mutants, and also for the higher order *glcat14* mutants in stem (**Figure 4-4D**).

Both Single and Double glcat14 Mutant Display Pleiotropic Growth Defects

Seeds obtained from WT and *glcat14* mutant lines were grown on MS plates supplemented with or without ABA to investigate the potential roles of GLCAT14A, B, and C in seed germination. The *glcat14* T-DNA mutant lines germinated like WT seeds under standard growth conditions. However, when the growth media was supplemented with 1 μM ABA, we observed that the *glcat14* mutants exhibited a delay in germination (**Figure 4-5**). Similarly, seeds grown under light conditions for 9 d showed no obvious differences in root growth relative to WT (**Appendix B: Supplemental Figure 4-2A**). Also, dark grown etiolated seedlings in both *glcat14* T-DNA mutant lines and the CRISPR lines showed no obvious differences in root growth relative to WT (**Appendix B: Supplemental Figure 4-2B**). Previous work reported increased length of hypocotyls and roots in *glcat14a* mutants of 5-day old etiolated seedlings (Knoch et al., 2013); however, it is noteworthy that our observation agrees with earlier work that showed no difference in hypocotyl length in etiolated seedlings except for *glcat14a/b/e* that were remarkably shorter (Lopez-Hernandez et al., 2020). We also investigated the role of GLCAT14A, B, and C in plant height, and our results showed that *glcat14a/c* and *glcat14b/c* and CRISPR lines (*glcat14a/b* and *glcat14a/b/c*) were significantly shorter than the WT plants (**Figure 4-4E and 4F**).

Figure 4-5

Germination Percentages of the glcat14 Mutants in the Presence of 1μM ABA

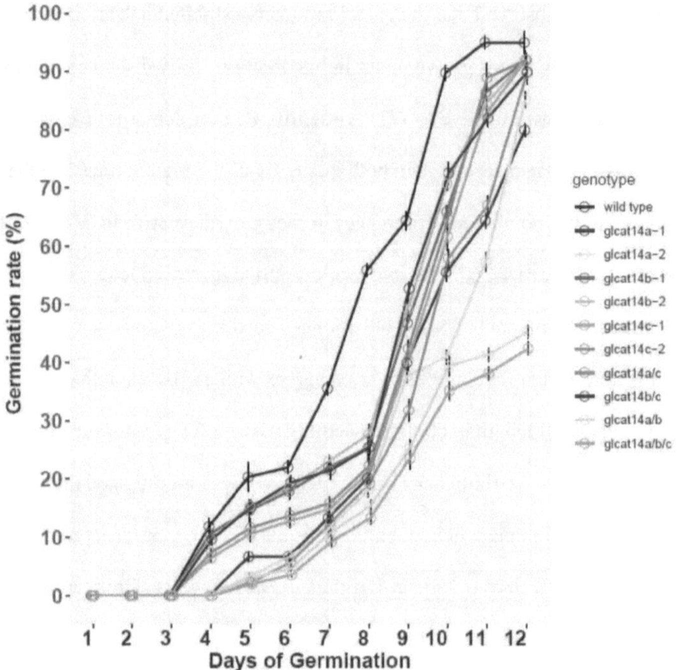

Note. Germination percentages of *glcat14* mutants showed delayed germination compared to WT in normal ½ MS supplemented with 1 μM ABA, which were grown for 12 days following 3 days of stratification at 4°C (See Materials and Methods). Twenty-five seeds of each genotype were used for germination, with three replicates.

Figure 4-6

Subcellular Localization of ATGLCAT14A, ATGLCAT14B and ATGLCAT14C

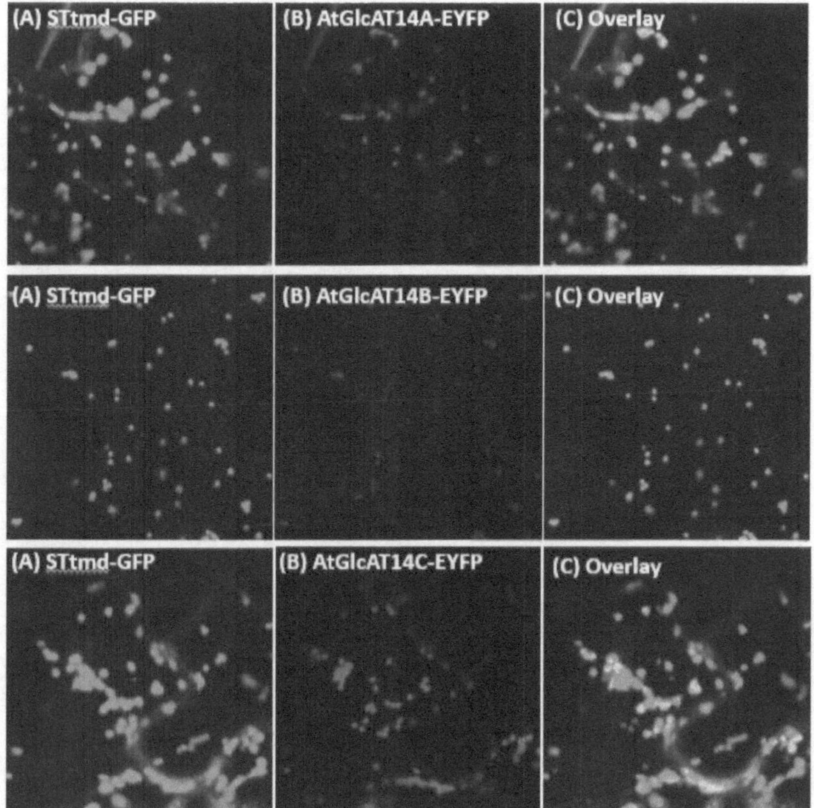

Note. Subcellular localizations of *ATGLCAT14A-EYFP, ATGLCAT14B-EYFP* and *ATGLCAT14C-EYFP*. STtmd-GFP represents sialyltransferase short cytoplasmic tail and single transmembrane domain fused to enhanced GFP, while *ATGLCAT14A, ATGLCAT14B* and *ATGLCAT14C* were fused to enhanced YFP to generate fusion protein, namely *ATGLCAT14A-EYFP, ATGLCAT14B-EYFP* and *ATGLCAT14C-EYFP*. Overlay in C, represents the overlaid image of panels A and panel B.

Previous work observed that CRISPR-Cas9 triple knockouts of *GLCAT14A*, *GLCAT14B* and *GLCAT14C* (Zhang et al., 2020) showed defects in trichome branching. A recent study similarly identified GLCAT14D and GLCAT14E to be involved in ensuring normal trichome development; this study also found that trichome defects in *glcat14* mutants were suppressed in a Ca^{2+} concentration dependent manner (Lopez-Hernandez et al., 2020). Using SEM, we examined the trichome defects of *glcat14a/b* and *glcat14a/b/c* and conducted an elemental composition analysis directed at estimating the amount of calcium using Energy Dispersive X-ray (EDX) microanalysis technique (Scimeca et al., 2018). Energy Dispersive X-ray (EDX) is an analytical technique commonly used to give semiquantitative measurements of surface elements within the surface layer of 1–2 microns in thickness (Thimmaiah et al., 2019). While the significant majority of the trichomes in *glcat14a/b* and *glcat14a/b/c* are one branched instead of the two branched trichomes characteristics of WT (**Figure 4-7A; Table 4-1**), EDX Spectra of investigated genotypes for elemental calcium showed no significant difference in the normalized mass of calcium relative to WT (**Figure 4-7D**). It is noteworthy to mention that none of the single and double *glcat14* T-DNA insertion mutants had defects in trichome branching as the trichome defects were only observed in *glcat14a/b* and *glcat14a/b/c* mutants.

Figure 4-7

SEM and SEM/EDX Elemental Calcium Analyses of Trichomes of WT, glcat14a/b, and

glcat14a/b/c Mutants

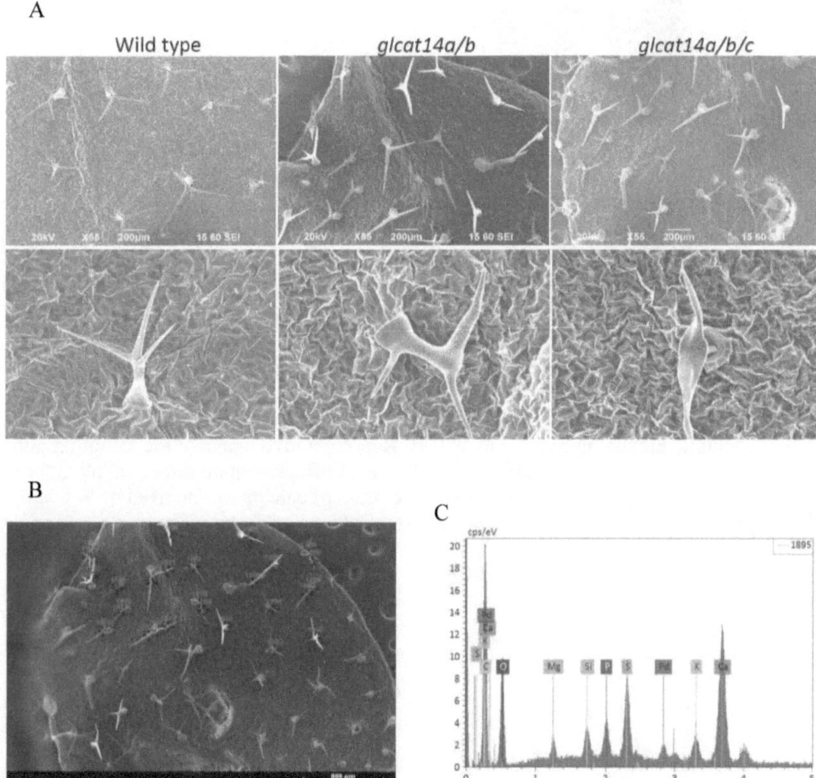

Figure 4-7: continued

D

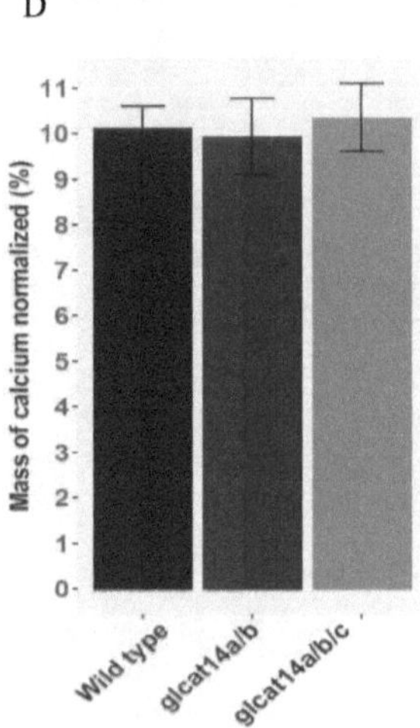

Note. A, SEM images of *glcat14a/b* and *glcat14a/b/c* mutants showing defects in trichome branching relative to WT. B, Representative image of the sampling points for SEM/EDX elemental analysis of trichomes. C, Representative spectral plots for the elemental composition of trichomes. D, Mass of calcium normalized of WT and *glcat14a/b* and *glcat14a/b/c* mutants showed no significant difference between WT and the *glcat14* mutants.

Table 4-1

Percentage of Trichome Branches in WT, glcat14a/b and glcat14a/b/c Mutants

Genotype	Number of Trichome branches			Total
	1	2	3	
Wild type	3.19	92.6	4.26	94
glcat14a/b	72.06	26.47	1.47	68
glcat14a/b/c	90.2	7.84	1.96	51

glcat14a/c Mutants Had Reduced Pollen Germination and Misshaped pollen

Scanning electron micrographs and *in vitro* pollen germination experiments were conducted to examine the pollen morphology and pollen germination of various *glcat14* T-DNA mutants. Previous work showed that *glcat14a/b* and *glcat14a/b/c* mutants displayed significant increases in defective pollen with concomitant effects on pollen germination (Zhang et al., 2020). In this study, we observed that in addition to *glcat14a/b* and *glcat14a/b/c* mutants, pollen from the *glcat14a/c* mutant also demonstrated a significantly lower germination rate compared to WT (**Figure 4-8B**), which could be reflective of the significant increase in defective pollen in *glcat14a/c* mutants in both the *in vitro* pollen germination (**Figure 4-8A and 8C**) and SEM (**Figure 4-8E**). Although, the pollen tube lengths of all *glcat14* mutants were comparable to WT (**Figure 4-8D**), an observation that is consistent with earlier studies (Zhang et al., 2020), some pollen tubes of *glcat14a/c* mutants displayed tip swelling relative to WT (**Figure 4-8A**) coupled with slight alterations in the reticulate pattern of Arabidopsis pollen (**Figure 4-8E**). The role of AGPs in pollen germination was demonstrated in previous work on *agp6* and *agp11* and *agp6/11* double mutants, which displayed reduced pollen germination (Coimbra et al, 2009), while another report in *Torenia fournieri* demonstrated that a disaccharide sugar, β-methyl-glucuronosyl galactose (4-Me-GlcA-β-1,6-Gal). present on AGPs, makes pollen tubes competent for ovule targeting/guidance (Mizukami et al., 2016). Considering that large amounts of pollen grains are produced during Arabidopsis sexual reproduction, there appears to be a significant proportion that are competent to ensure fertilization, given that the silique lengths and seed set in *glcat14a/c* mutants were comparable to WT (data not shown).

Figure 4-8

Pollen Phenotypes of Selected glcat14 Mutants

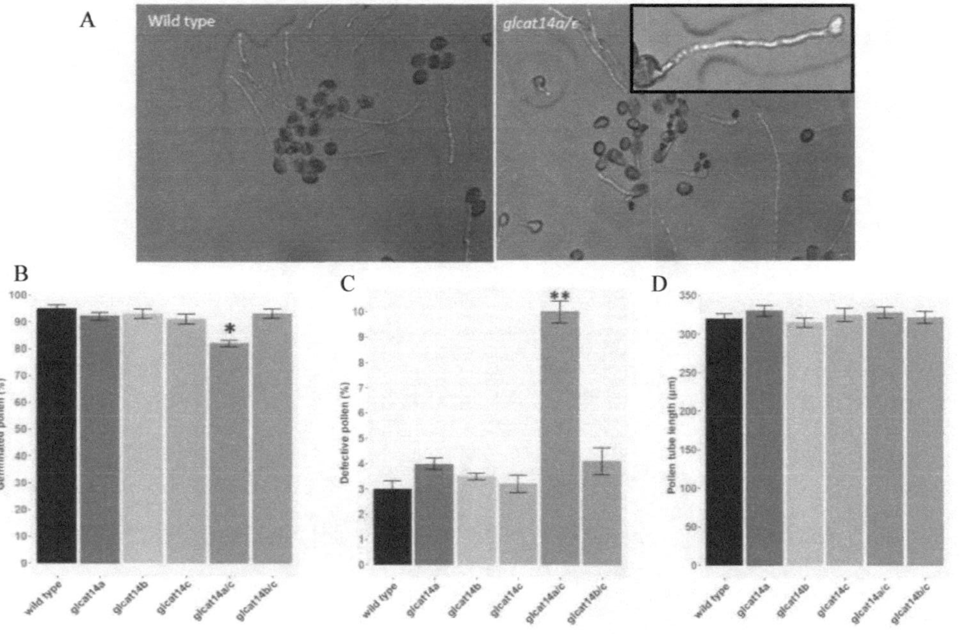

E

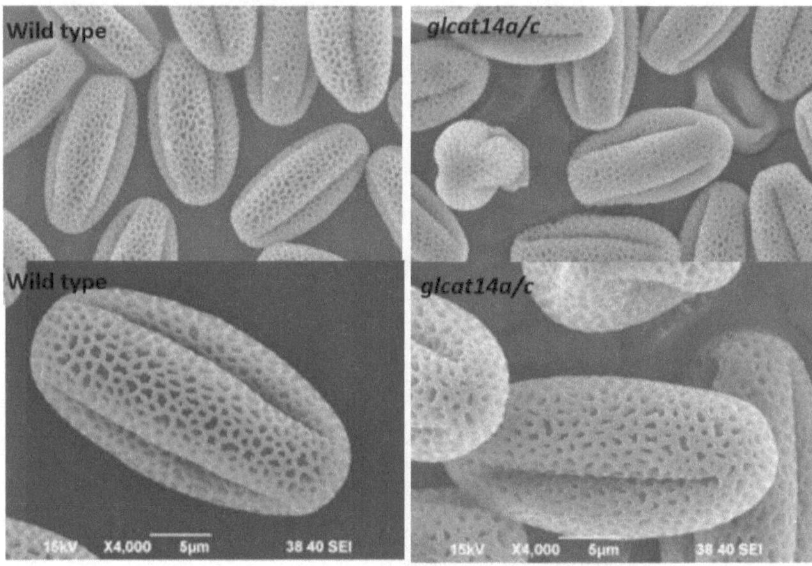

Note. A, *In vitro* pollen germination showed defective pollen (arrowheads) and pollen tube tip swelling (arrow; inset in A) in the *glcat14a/c* mutant compared to WT. B, Pollen germination (%) in *glcat14* mutants and WT showed decreased pollen germination in *glcat14a/c* mutants relative to WT. C, Proportion of defective pollen were significantly higher in *glcat14a/c* mutants relative to WT. D, Pollen tube lengths were similar in *glcat14* mutants and WT. All measurements were taken 3 h after incubation of pollen grains on pollen germination media. Approximately 150 pollen grains were measured for each genotype with three replicates. (* $P < 0.05$; ** $P < 0.01$); E, SEM images of *glcat14a/c* mutants showed misshaped pollen and slight alterations in the reticulate structure relative to WT.

Discussion

A large number of Arabidopsis genes are predicted to encode glycosyltransferases that are involved in cell wall biosynthesis (Shimoda et al., 2014). Despite these predictions, very few candidate genes have been functionally characterized. Genetic strategies deployed to elucidate glycosyltransferase functions have been hampered by functional redundancy, as many genes have been identified to play similar roles in the same and/or different organs/tissues. As a result, multiple genes have to be disrupted before biochemical or physiological roles of these cell wall genes can be deduced. Previous work identified eleven *GLCAT* genes in Arabidopsis that belong to the CAZy GT14 family (Ajayi and Showalter, 2020; Ye et al., 2011); these genes may be involved in the glucuronidation of Type II AGs, given their conserved genetic architecture and the conserved GLCAT domain present in genes (Ajayi and Showalter, 2020).

While the functional roles of some *GLCAT* genes have been identified (i.e., *GLCAT14A-E*; **Figure 4-1B**), the functional roles of the remaining putative *GLCATs* remain to be verified and studied further. Work done previously using CRISPR-Cas9 knockouts of *GLCAT14A-C* showed severe pleiotropic growth defects especially in the higher order mutants (Zhang et al., 2020), while subsequent work by another group extended this work and identified developmental defects in additional Arabidopsis *glcat14* mutants that were partially suppressed by supplementing the growth medium with calcium (Lopez-Hernandez et al., 2020). In this report, T-DNA insertion mutants and CRISPR lines of *GLCAT14A-C* were utilized to corroborate previous findings and to obtain additional insights into the biochemical and physiological functions of

GLCAT14A-C in Arabidopsis. Specifically, we sought to understand the degree to which the alterations in the amount of GlcA and calcium content accounts for the mutant phenotypes using biochemical and molecular genetic approaches.

Plant cell walls are important dynamic structures that need to be monitored properly to ensure their functional integrity (Vaahtera et al., 2019). Alterations in the polymer structure often trigger compensatory changes in the cell wall during normal growth or in changing environments. Strong support for the continuous cellular surveillance mechanism aimed at optimizing plant cell function exist (Hamann and Denness, 2011). The cell wall integrity maintenance mechanism constantly monitors and perceives changes in biological processes and coordinates activities to promote plant growth through restructuring of glycan moieties (Hamann, 2015; Basu et al., 2015). In order to examine whether there are alterations in AGP abundance in *glcat14* mutants relative to WT, we quantified the amount of AGPs in rosette leaves, stems, and siliques using β-Gal Yariv reagent that selectively binds to AGPs (Kitazawa et al., 2013). We observed that the amount of glycosylated AGPs were altered in the various organs of the *glcat14* mutants, with increased amounts of Yariv precipitable AGP detected in siliques and stems compared to rosette leaves (**Figure 4-3**). Moreover, an increase in AGP content was detected in *glcat14a, glcat14a/c, glcat14a/b* and *glcat14a/b/c* mutants in the rosette leaves and stems, while in the siliques, *glcat14a, glcat14c, glcat14a/c, glcat14a/b* and *glcat14a/b/c* mutants showed an increase in AGP content. Although previous work reported an increase in Yariv precipitable AGPs in CRISPR-Cas9 generated *glcat14ab* and *glcat14abc* (Zhang et al., 2020), our findings indicate that the genetic knock out of

GLCAT14A plays a role in the increased Yariv precipitable AGP content observed for *glcat14a/c*, *glcat14a/b* and *glcat14a/b/c* mutants. In addition, the increase in Yariv precipitable AGP observed among *glcat14* mutants relative to the WT might be a compensatory response that is necessary for maintaining cell wall integrity and establishing normal plant physiological functions.

Considering the observed alterations in AGP content across the investigated organs in the *glcat14* mutants, we conducted monosaccharide composition and western blot analyses to gain additional insights into changes in the glycosylation processes in these mutants. We observed organ-specific glycosylation responses, reflected by organ-specific alterations in the galactose content in *glcat14* mutants investigated. Previous work on the sugar analysis of AGP extracts of the aerial parts of *glcat14* mutants showed an increase in the galactose content in *glcat14ab* and *glcat14abc* (Zhang et al., 2020), but our findings indicate that this pattern is not the same for all organs. While we observed an increase in galactose contents in the leaves (**Figure 4-4A**) and siliques (**Appendix A: Supplemental Table 4-3**) for *glcat14a/b* and *glcat14a/b/c*, we observed a significant reduction in galactose content in the stem for *glcat14a/b/c* mutants relative to WT (**Figure 4-4C**). Overall, the observed increase in galactose content in the *glcat14* T-DNA mutant and CRISPR mutant lines (**Appendix A: Supplemental Table 4-3-5A**) centers around the idea that GlcA residues serve as a "cap" that can terminate the elongation of galactan chains and hence the absence of such GlcA residues could lead to an increase in the length of galactan chains (Knoch et al., 2013); however, this idea was challenged elsewhere (Lopez-Hernandez et al., 2020). Our data indicates that organ-specific cell wall changes do occur in response to structural changes, to ensure that plant cell walls can perform their biological functions in different situations (Basu et al., 2015).

Immunolabelling of β-GlcAs of Type II AGs showed weak immunolabelling with LM2, an antibody that targets β-GlcAs of Type II AGs, in *glcat14a/b* and *glcat14a/b/c* stem and leaf extracts compared to the WT which had increased LM2 binding. This finding corroborates the observed reduction of GlcA in the monosaccharide composition analysis, supporting the argument that other GLCATs may be involved in the AG glucuronidation processes. The LM2 signal was detected in the high molecular mass region (**Figure 4-2A and 2B**), which might be a due to AGPs interaction with other cell wall polymers as demonstrated in APAP1 mutants (Tan et al., 2013). We observed weak binding of LM2 antibody in stem compared to the leaf, and this differences in LM2 binding may be due to the micro-heterogeneity of the AGP glycan in stem and leaf.

Arabidopsis *GLCAT14A-C* are highly expressed during seed germination (Ajayi and Showalter, 2020). Moreover, *GLCAT14A* and *GLCAT14B* are primarily expressed in the micropylar endosperm during seed development (Winter et al., 2007). We observed significant delays in seed germination in the presence of 1 μm ABA in *glcat14* mutants relative to WT (**Figure 4-5**), a finding that is consistent with an earlier report (Zhang et al., 2020). ABA is reported to influence the cytosolic calcium oscillations and impact many physiological processes like seed germination and environmental stress responses (Guo et al., 2002; Finkelstein et al., 2002) and such oscillations in cytoplasmic calcium concentrations could be mediated by calcium that is bound by GlcA residues in AGPs (Lamport, 2013). In any case, further research is needed to verify this idea.

Following the localization of ATGLCAT14A in the Golgi apparatus (Knoch et al., 2013), we investigated the subcellular localizations of ATGLCAT14B and ATGLCAT14C. We observed that ATGLCAT14A-EYFP and ATGLCAT14B-EYFP colocalized with the STtmd–GFP, except for ATGLCAT14C-EYFP which partially colocalized with STtmd–GFP (**Figure 4-6A-C**). The detection of ATGLCAT14A, ATGLCAT14B and ATGLCAT14C as punctate vesicles were consistent with the idea that ATGLCAT14A, ATGLCAT14B and ATGLCAT14C function in the glucuronidation of AGPs at unique positions in the type II glycan structure (Lopez-Hernandez et al., 2020), and therefore was expected to be localized to the Golgi.

The importance of *GLCATs* in leaf development is reflected by the significant reduction in GlcA content in *glcat14a/b* and *glcat14a/b/c* mutants relative to WT as revealed by monosaccharide composition and western blot analyses (**Figure 4-2A and 4A**). Using SEM, we observed defects in trichome branching in *glcat14a/b* and *glcat14a/b/c* mutants (**Figure 4-7A**), and this finding is consistent with earlier work (Lopez-Hernandez et al., 2020; Zhang et al., 2020). While most of the *glcat14a/b* and *glcat14a/b/c* trichomes were single branched trichomes (**Table 4-1**), a much more severe trichome branching defect was reported for *glcat14a/b/e* mutants which could be partially suppressed by supplementing the growth media with calcium (Lopez-Hernandez et al., 2020). Energy Dispersive X-ray (EDX) analysis is an analytical technique commonly used to provide semi quantitative measurements of surface elements within a surface layer of 1–2 microns in thickness (Coimbra et al., 2009).

We utilized this technique to estimate whether there were differences in the amount of calcium in trichomes of *glcat14a/b* and *glcat14a/b/c* mutants. We found that the normalized calcium mass of trichomes of *glcat14a/b* and *glcat14a/b/c* mutants were comparable to WT based on SEM-EDX technique (**Figure 4-7D**). Similarly, the calcium content of leaf AGP extracts obtained from *glcat14a/b* and *glcat14a/b/c* mutants were similar to WT (**Figure 4-4B**). Despite this finding, we cannot exclude that calcium plays a role in establishing normal trichome branching and may be involved during the earlier stages of trichome development as demonstrated earlier (Lopez-Hernandez et al., 2020). It is worth mentioning that most GlcA residues in leaf AGPs are methylated (Tryfona et al., 2010), and it remains to be determined whether certain GlcA residues added by particular GLCATs are preferentially methylated or involved in calcium binding and how such modifications impact trichome development in Arabidopsis.

This investigation in understanding the roles of GLCAT14A, B, and C in plant growth and development identified significant reductions in plant height in *glcat14a/c*, *glcat14b/c* T-DNA insertion mutant lines relative to WT (**Figure 4-4E and 4F**). Previous work showed that *glcat14bc*, *glcat14ab* and *glcat14abc* were shorter than WT (Zhang et al., 2020), while severe growth defects were also reported for *glcat14a/b/d* and *glcat14a/b/e* (Lopez-Hernandez et al., 2020). We found that *glcat14a/c* mutants were also significantly shorter than the WT but were not as severe as those observed in *glcat14a/b* and *glcat14a/b/c* mutants (**Figure 4-4E and 4F**). The importance of calcium was recently demonstrated by the observed hypersensitivity of the inflorescence stem lengths of some *glcat14* mutants to low calcium concentrations (Lopez-Hernandez et al., 2020). In addition, the significant reductions in calcium were only observed in the stem, while the calcium contents of *glcat14* mutants in leaves and siliques were comparable to WT. Considering the complex framework of plant cell wall glycans, the maintenance of cell wall structure and expansion may not be limited to the interaction of Ca^{2+} with AGP and/or pectins, as other possible interactions of calcium may exist with cell wall modifying enzymes (Hepler, 2005; Vanneste and Friml, 2013).

Chapter 5: Glucuronidation of Type II Arabinogalactan Polysaccharides Function in Sexual Reproduction of Arabidopsis

Introduction

Double fertilization is a unique reproductive system shared by all angiosperms for the successful development of viable offspring. In Arabidopsis, the male reproductive organ comprises of the stamen, which consists of an anther and a stalk-like filament, which aids pollen adherence to the stigma (Scott et al., 2004). Several developmental events are initiated during pollen development as microsporocytes undergo meiosis to form tetrads and then free microspores that generates mature pollen after mitosis (McCormick, 2004). Following the attachment of mature pollen to stigmatic surfaces, the pollen tube grows toward the ovule through the stigmatic cells and the transmitting tract in style and are lured along the funiculus into the micropyle of the ovule by chemoattractant such as AtLURE1, a cysteine-rich peptide that is required and sufficient for pollen tube attraction in *A. thaliana.* (Takeuchi and Higashiyama, 2012). The pollen tube penetrates the degenerated synergid and then bursts to release two sperm cells into a narrow receptive region between the egg cell and the central cell, and communication between the four gametes (namely the egg cell, the central cell, and two sperm cells) is initiated to ensure reproductive success (Hamamura et al., 2011; Huang et al., 2015). The intercellular communications between the male and the female gametes is a complex process that involves the proper coordination of signals and several molecules for successful fertilization (Mizuta and Higashiyama, 2018; Zhong and Qu 2019; Duan et al., 2020).

Arabinogalactan proteins (AGPs) are complex hyperglycosylated cell wall proteoglycans that function in plant sexual reproduction (Coimbra et al., 2007). Arabinogalactan proteins are abundant in reproductive tissues, and may function in gametophytic cell differentiation, pollination and pollen tube growth (Cheung and Wu, 1999), egg cell fertilization, and zygote division (Qin and Zhao, 2006). AGPs are glycosylated by several glycosyltransferases and antibody binding specific to AGP glycan moieties have been used to elucidate the precise function of AGP glycans in anther and ovule development (Coimbra et al., 2007). In addition, several studies demonstrated the importance of AGPs in male and female gametophyte development in angiosperms. For example, over-expression of *BcMF18* in Arabidopsis and *FLA3*, a fasciclin-like arabinogalactan protein, affects microspore development and impacts pollen wall intine formation (Li et al., 2010; Lin et al., 2019). Similarly, *AtAGP6* and *AtAGP11*, have been implicated in pollen development (Coimbra et al., 2009) and loss of function of AtAGP6 and AtAGP11 showed collapsed pollen grains, reduced (and precocious) pollen grain germination and tube elongation, with concomitant effects resulting in shorter fruits with less seeds (Coimbra et al., 2009, 2010). In addition, loss of function of *KNS4/UPEX1*, a member of the CAZy glycosyltransferase 31 gene family involved in the galactosylation of AGP protein cores, resulted in the abnormal development of the primexine matrix laid on the surface of developing microspores (Dobritsa et al., 2011; Suzuki et al., 2017). The multi-layered pollen wall is made up of an inner pectocellulosic intine and an outer sporopollenin-based exine (Scott et al., 2004). The exine is further subdivided into inner nexine and outer sexine. The sexine has a three-dimensional (3D) structure composed of many columns called baculae and a roof called tectum and their combinations give rise to the reticulate arrangements with a uniform mesh size (Suzuki et al., 2017). Both the sporophytic tapetum and the male gametophyte are regulators responsible for exine synthesis (Schnurr et al., 2006), while the intine synthesized beneath the exine, is controlled by the microspore (Schnurr et al., 2006; Ariizumi and Toriyama 2011).

During sexual reproduction, several genes/proteins have been identified to be involved in ensuring proper sperm delivery and fusion to the egg cell and the central cell. The fertilization-independent seed (FIS)-class Polycomb Repressive Complex 2 (FIS-PRC2) have been implicated in polytubey block through FIS-PRC2 pathway (Maruyama et al., 2013). Maruyama et al. (2015) showed that the egg and the central cell independently but coordinately control polytubey block through the rapid dilution of pollen tube attractants initiated by the synergid-endosperm fusion. The only AGP implicated in polytubey block is AGP4 and loss of function of AGP4, named JAGGER, displayed increased rate of polytubey due to the failure of the persistent synergids to degenerate (Pereira et al., 2016). Arabidopsis fail to block polytubey when the egg cell or central cell remain unfertilized, allowing the next pollen tube to recover the early fertilization failure (Beale et al., 2012; Kasahara et al., 2012; Maruyama et al., 2013).

Despite the involvement of some AGPs in ensuring proper development of male and female gametophytes, reports on the biological role of AGP glycan moieties in Arabidopsis sexual reproduction are rare. Previous work showed that *glcat14a glcat14b* and *glcat14a glcat14b glcat14c* mutants displayed defective pollen with shorter siliques and low seed set (Zhang et al., 2020), but the contributing factors are unknown. Also, a similar observation was made for *galt2 galt3 galt4 galt5 galt6* quintuple mutants, which showed defective pollen with shorter siliques and seed set (Zhang et al., 2021), suggesting that both galactosylation and glucuronidation of AGP protein cores are critical to Arabidopsis sexual reproduction. Here, using biochemical and immunolabelling techniques, we extended the work on three glucuronosyltransferase (*GLCAT14A*, *GLCAT14B* and *GLCAT14C*) genes, involved in the addition of glucuronic acid to type II AGs, and elucidated their roles in Arabidopsis sexual reproduction. Specifically, we showed that the loss of function of *GLCAT14A*, *GLCAT14B* and *GLCAT14C* resulted in the poor organization of the reticulate structure of the exine wall, with abnormal development of the intine layer, leading to large number of non-viable pollen grains in *glcat14a/b* and *glcat14a/b/c* mutants. Also, synchronous development between locules within the same anther was lost in some *glcat14a/b/c* stamens, while some anther locule lacked pollen grains. In addition, we showed that GLCAT14A, GLCAT14B and GLCAT14C are important in polytubey block and are critically important in ensuring normal embryo development and seed set in Arabidopsis. Our data reveal not only the important role AG glucuronidation plays in establishing normal microspore development but also highlights its potential role in crosstalk involving male and female gametophytes.

Materials and Methods

Source and Plant Growth Conditions

Arabidopsis thaliana (Columbia ecotype) was used as the WT and was obtained from the Arabidopsis Biological Research Center (ABRC), Columbus, Ohio, USA. The

GLCAT14A-AT5g39990, *GLCAT14B-AT5g15050*, and *GLCAT14C-AT2g37585* genes were generated using CRISPR-Cas9 multiplexing approach. All the guide RNA design, vector construction and indel confirmation by sequencing have been described previously (Zhang et al., 2020). For the purpose of this study, the CRISPR-Cas9 knock outs of GLCAT14A and GLCAT14B are referred to as *glcat14a/b* mutants while CRISPR-Cas9 knock outs of GLCAT14A, GLCAT14B and GLCAT14C are referred to as *glcat14a/b/c* mutants. All plants (WT, *glcat14a/b* and *glcat14a/b/c*) used in this study were germinated after 4 days of stratification in the dark at 4 °C and were grown under long-day conditions (16 h of light/8 h of dark, 22 °C, 60 % humidity) in growth chambers.

In-vitro Pollen Germination and Pollen Tube Measurements

Matured flowers from 35-day-old plants of WT, *glcat14a/b* and *glcat14a/b/c* mutants were used for an *in vitro* pollen germination assay following methods described previously (Zhang et al., 2020). Briefly, mature pollen grains were grown on a germination medium containing 10% sucrose, 0.01% boric acid, 1mM $CaCl_2$, 1mM $Ca(NO_3)_2$, 1mM KCl, 0.03% casein enzymatic hydrolysate, 0.01% myo-inositol, 0.1mM spermidine, 10mM GABA, 500 µM methyl jasmonate, and 1% low-melting agarose. Pollen tube lengths were measured 3 h after incubation with a Nikon phot-lab2 microscope at 20x magnification. Around 200 pollen tubes were measured in an individual experiment with three replicates. Pollen tube measurements were carried out using imageJ and *glcat14* mutants were compared to WT.

Electron Microscopy

For SEM observations, pollen grains and anther of WT, *glcat14a/b* and *glcat14a/b/c* were mounted on aluminum stubs using double adhesive tapestubs and sputter coated with a palladium alloy using a Cressington 208C high-resolution sputter coater (Ted Pella Inc.) at the Institute for Corrosion and Multiphase Technology, Ohio University. Images were captured using a JEOL JSM-6390 scanning electron microscope (Hitachi High-Technologies). For TEM observations, ultrathin sections of resin-embedded anthers were prepared at the Ohio State University using an ULTRACUT N ultramicrotome (Reichert-Nissei, Tokyo, Japan) with a diamond knife and sections were mounted on copper grids essentially as described by Suzuki et al. (2008). Specimens were viewed by FEI TEcnai G2 Spirit TEM using the Campus Microscopy and Imaging Facility at the Ohio State University. Measurements of exine thickness of developing microspores were carried out using imageJ and statistical comparisons were made between WT and *glcat14* mutants.

Immunolabelling and Microscopy

Following the *in-vitro* pollen germination of WT and *glcat14* mutants for 6hrs, pollen tube in pollen germination medium were fixed following methods described earlier (Dardelle et al., 2010). After fixation, PTs were washed three times with PIPES buffer and then three times with phosphate-buffered saline (PBS) + 3% bovine serum albumin (BSA). Immunolabelling with JIM5 and LM2 primary antibodies (Knox *et al.,* 1990) were diluted at 1:5 as described previously (Mollet et al., 2002) with PBS in 1% BSA. Pollen tubes were rinsed with the buffer (PBS in 1% BSA) and incubated overnight at 4°C. The primary antibodies were removed and washed three times in buffer. Anti-rat fluorescein isothiocyanate (FITC)-conjugated secondary antibody (diluted 1:100 in PBS in 1% BSA) was added, and the tubes were incubated at room temperature in the dark for 2 h. After incubation, PTs were washed three times with PBS followed by three washes with DI water. Controls were carried out by incubation of the pollen tubes with the secondary antibody only. For Ovule AGP and pectin detection, ovules from naturally pollinated pistils of flower buds at stage 12 and stage 14 (Smyth et al., 1990) were transferred to fixative as described previously (Duan et al., 2020) and processed for pectin and AGP detection using JIM5 and JIM13 primary antibody, respectively, and subsequently incubated with FITC-labelled secondary antibody against rat antibodies respectively, as described previously (Duan et al., 2020). Imaging of processed samples were carried out by epifluorescence microscope using the FITC filter. Control experiments were performed by omitting the incubation with the primary antibody (incubation with blocking solution only) and showed no unspecific staining. For immunolabelling of sectioned samples, 1-2 μm of LR White embedded anthers were immunolabelled with JIM5, JIM7 and JIM13. JIM5 and JIM7, respectively, detect partially methylesterified homogalacturonan epitope (about 40% unesterified residues adjacent to, or flanked by, residues with methylester groups) and partially methylesterified homogalacturonan epitope (methylesterified residues up to 80%) (Clausen et al., 2003), while JIM13 recognizes AGPs (Knox et al., 1990). The immunolabelling of sections were carried out essentially following methods described previously (Suzuki et al., 2017) and imaging of processed samples were carried out by epifluorescence microscope using the FITC filter.

Staining of Pollen Grains and Ovule

Ruthenium red was used to detect the presence of pectin in mature pollen grains and ovules (stage 13-16) of WT and *glcat14a/b/c* mutants (Smyth et al., 1990). For vanillin staining, siliques were collected from WT and *glcat14a/b* and *glcat14a/b/c* mutants, three day after pollination and stained with 1% vanillin (4-hydroxy-3-methoxybenzaldehyde) in 6 N HCl (Sigma) without dissection. Alexander's solution was used to test the pollen viability (Alexander, 1969). To observe the nuclei and callose wall, pollen grains were stained in DAPI solution (Regan and Moffatt, 1990) and 0.1% (w/v) aniline blue in 0.1 mol L^{-1} K_2HPO_4–KOH buffer (pH 11) respectively. Aniline blue (AB) staining between WT and *glcat14a/b/c* mutants were conducted to test *in vivo* pollen tube growth towards the ovule. Pistils at 24 h after pollination (HAP) were excised and processed following methods described previously for AB staining (Mori et al., 2006) and visualized using an epifluorescence microscope. For toluidine blue staining, sections (1μM) of resin embedded anthers were mounted on a glass slide and stained as described previously (Suzuki et al., 2017).

Monosaccharide Composition Analysis by High Performance Anion Exchange Chromatography with Pulsed Amperometric Detection (HPAE-PAD)

AGPs were isolated from Arabidopsis flowers of WT, *glcat14a/b* and *glcat14a/b/c* mutants following methods described earlier (Lamport, 2013). Fifty microliters of 10mg/mL AGP were hydrolyzed using 2 N TFA at 121°C for 90 min. TFA was removed by evaporation with N_2 gas. Samples were dissolved in 500 μL milli-Q water containing 0.2mM cellobiose as an internal standard. A standard sugar mixture (fucose, rhamnose, arabinose, galactose, glucose, xylose, mannose, galacturonic acid, and glucuronic acid) was used for making the standard curve. Monosaccharide compositions were calculated as molar percentages (mol %). All samples and standards were subjected to high pH anion-exchange chromatography with pulsed amperometric detection (HPAE-PAD) using a Dionex PA-20 column (Thermo Fisher Scientific, Sunnyvale, CA, USA) essentially as described here (Obro et al., 2004).

Western Blotting Analysis

The total protein extraction from Arabidopsis flowers of WT, *glcat14a/b* and *glcat14a/b/c* mutants were performed following methods described previously (Fragkostefanakis et al., 2012). Total protein supernatants from the samples extracted were quantified by the Bradford method (Bradford, 1976). Twenty-five micrograms (25μg) of crude protein extract of WT, *glcat14a/b* and *glcat14a/b/c* mutants were separated by 10% mini-PROTEAN® TGX™ (456-1033; Bio-Rad, **https://www.bio-rad.com**) gel electrophoresis and transferred onto a 0.45-μm Immun-Blot low florescence PVDF membrane (1620260; Bio-Rad) using wet transfer system at constant current of 100V for 45 mins. The blots were blocked for 1 h (3% BSA, in 1× Tris-Buffered Saline with Tween [150 mM NaCl, 20 mM Tris base, 0.1% Tween 20; TBST]), probed with 1:10 AGP specific LM2 and JIM13 monoclonal primary antibody (Carbosource, **https://www.ccrc.uga.edu**) in blocking buffer for 1 hr, washed three times with 1× TBST for 15 min, probed with goat anti-rat IgG H + L secondary antibody

(PI31629; Fisher Scientific), washed thrice with $1\times$ TBST for 10 min, once with $1\times$ TBS for 10 min, treated with chemiluminescent substrate (Clarity ™ ECL substrate, 1705060S; Biorad) and ChemiDoc imaging system (Bio-Rad) was used for image acquisition. PVDF membranes were incubated for 10 min with Ponceau S solution [0.1% (w/v) Ponceau S; 5% (v/v) acetic acid] to stain for total proteins.

Evaluation of Seed Set, Silique Measurements and Embryogenesis

Siliques were collected from 40-day-old WT and *glcat14* mutants for measurement of silique lengths and seed set. Siliques corresponding to each stage of embryo development (Le et al., 2010) were harvested and cleared in Hoyer's solution essentially as described previously (Bayer et al., 2009, Le et al., 2010) and imaged using an epifluorescence microscope.

Results

glcat14a/b and glcat14a/b/c Mutants Showed Defects in Reticulate Structure of Tectum and Reduced Viable Pollen

A mutant phenotype previously identified in *glcat14a/b* and *glcat14a/b/c* mutants involving Arabidopsis sexual reproduction was defective pollen grain that resulted in reduced pollen germination (Zhang et al., 2020). However, the underlying reasons for these observations was not investigated and was lacking. We investigated the pollen grain morphology using SEM and we observed that in contrast to the WT, the reticulate network in *glcat14a/b* and *glcat14a/b/c* were completely distorted with defective exine formation (**Figure 5-1D-1F**). The lacunae, which is described as the hole within the tectum, including the space directly beneath the tectum hole (Suzuki et al., 2017), were either reduced or blocked in most of the *glcat14a/b* mutant in comparison to the WT, while most of the tectum in *glcat14a/b/c* mutants were partially broken, and in some cases, the bacula were exposed (**Figure 5-1D, 1E, 1F**). Also, we observed that a greater percentage of the pollen grain area in the WT were between 300-450 µM in size while more than 40% of *glcat14a/b* and *glcat14a/b/c* pollen grains were between 100-300 µM in size (**Figure 5-1J**). In addition, the anther in *glcat14a/b* and *glcat14a/b/c* mutants were more compacted relative to the WT (**Figure 5-1G-1I**). Notably, the reticulate structure in *glcat14a*, *glcat14b*, *glcat14c*, and *glcat14b/c* and were comparable to WT, except for the small lacunae that was present in *glcat14a/c* mutants (**Appendix B: Supplemental Figure 5-1D-1I**). Also, we observed the fusion of pollen grains through their exine walls (**Arrow in Appendix B: Supplemental Figure 5-1B, 1C, inset in 1C**), while Aniline blue staining showed the attachment of some pollen grains to cell wall debris in *glcat14a/b* and *glcat14a/b/c* mutants (**Figure 5-2B and 2C**). Given the observations using SEM, we carried out Alexander staining to examine the proportion of viable and non-viable pollen and we observed a higher proportion of non-viable pollen in both *glcat14a/b* and *glcat14a/b/c* mutants relative to the WT (**Figure 5-2D-2F; Table 5-1**), and this observation was consistent with Ruthenium red (RuR) staining of pollen grains (**Figure 5-2G-2I**). Interestingly, we observed the presence of sperm nuclei in most of the matured pollen stained with DAPI for *glcat14a/b* and *glcat14a/b/c* mutants despite defects in exine formation (**Figure 5-2J-2L**).

Figure 5-1

SEM Examination and Pollen Area of WT, glcat14a/b and glcat14a/b/c Mutants

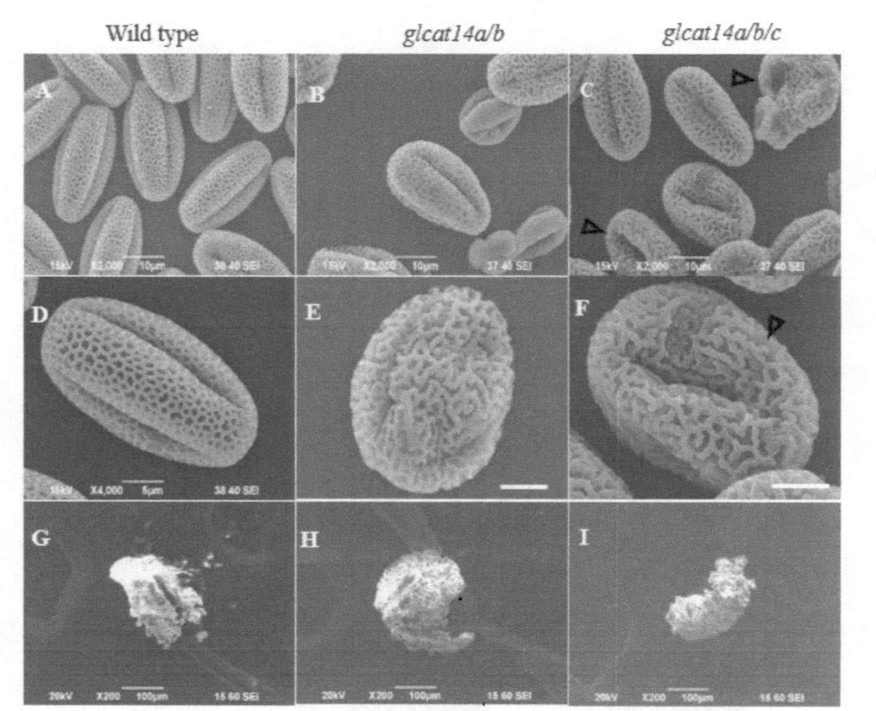

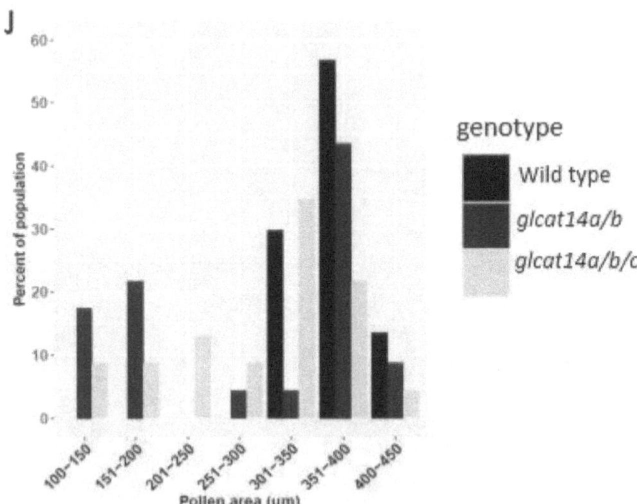

Note. SEM images of the surfaces of wild-type (A, D), *glcat14a/b* (B, E) and *glcat14a/b/c* (C, F) pollen grains. Several of the *glcat14a/b* and *glcat14a/b/c* pollen grains displayed abnormal exine patterns, with bacula exposed in some *glcat14a/b/c* pollen grains (arrowheads in C, F). SEM images of anthers from wild-type (G), *glcat14a/b* (H) and *glcat14a/b/c* (I) showed that pollen grains coalesce to prevent easy release of pollen grains in *glcat14* mutants relative to wild type. Proportion of pollen area in wild type, *glcat14a/b* and *glcat14a/b/c* mutants (J) showed that the majority of the pollen grains in *glcat14a/b* and *glcat14a/b/c* are smaller than in wild type. Scale bar in E and F = 5µm, while scale bar for other images are as indicated

Figure 5-2

Pollen Defects in Observed glcat14a/b and glcat14a/b/c Mutants

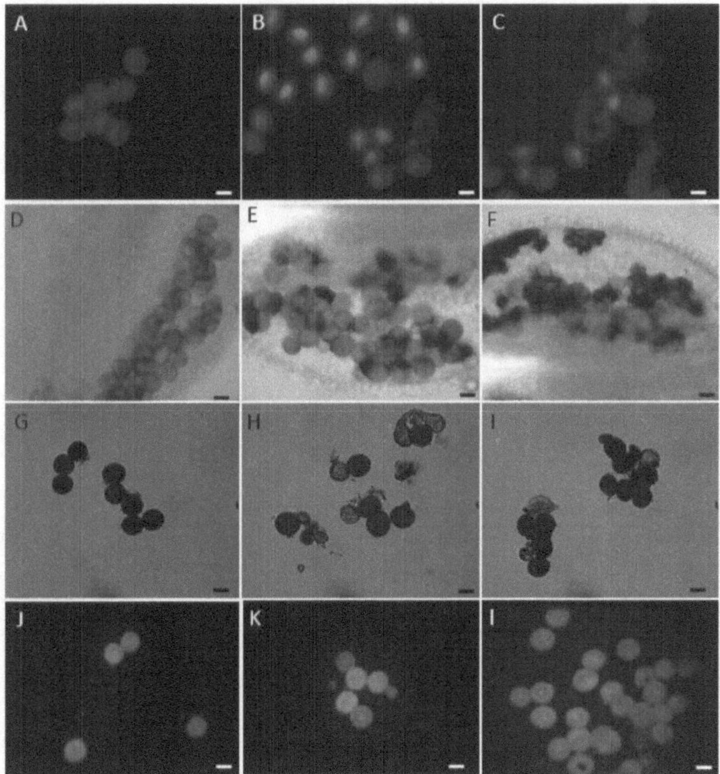

Note. Aniline blue staining of WT (A), *glcat14a/b* (B) and *glcat14a/b/c* mutants (C). Note the presence of non-viable pollen and cell wall debris (B, C). Alexander staining showed the presence of defective pollen (purple coloration) in *glcat14a/b* (E) and *glcat14a/b/c* mutants (F) in contrast to the WT (D). Ruthenium red staining that binds pectin showed reduced pectin in aborted pollen in *glcat14a/b* (H) and *glcat14a/b/c* mutants (I) relative to the WT (G). DAPI staining detected the presence of sperm nuclei in both WT (J) and *glcat14* mutants (K, L) with the exception of aborted pollen grains. Scale bar in A-L = 50μm.

Table 5-1

Proportion of Normal and Misshaped Pollen in WT, glcat14a/b and glcat14a/b/c mutants as revealed by SEM and Alexander Staining

	Number of normal and mishaped pollen[a]					
	Normal		Mishaped		Total[b]	
Genotype	Alexander staining	SEM	Alexander staining	SEM	Alexander staining	SEM
Wild type	98.2	99.4	1.8	0.58	123	694
glcat14a/b	59.2	55.8	40.8	44.2	79	488
glcat14a/b/c	51.5	55.4	48.5	44.6	89	456

[a] *Percentages of pollen having the indicated pollen morphology;* [b] *Represents the number of the total pollen stained with Alexander staining and SEM*

Microspore Abortion in glcat14a/b and glcat14a/b/c Mutants are due to Poorly
Developed Pollen Intine and Exine Wall

Because pollen wall patterning is being structured and constructed during pollen
development (Twell, 2010), we investigated pollen development by examining the
embedded sections stained with Toluidine O dye, and imaging of anther sections using
transmission electron microscope (TEM). We observed developmental abnormalities in
glcat14a/b and *glcat14a/b/c* mutants during microscope development. Specifically, we
found that the bacula and the tectum arrangements were disorganized and characterized by
poorly formed exine in *glcat14a/b* and *glcat14a/b/c* mutants relative to the WT (**Figure 5-
3A-3C**). Similarly, the number of bacula (counted only when the base of the baculae
touches the nexine along the circumference of the microspores) appeared reduced in
glcat14a/b/c mutants (Mean ± SD; WT, 58.7 ± 6.8 and *glcat14a/b/c* , 47.6 ± 8.3; n=10),
while the exine thickness, equivalent to the length of bacula (Suzuki et al., 2008) of
glcat14a/b and *glcat14a/b/c* mutants were comparable to WT (Mean ± SD; WT, 1.3 ± 0.15;
glcat14a/b, 1.32 ± 0.06 and *glcat14a/b/c*, 1.29 ± 0.13; n > 85 per genotype). Interestingly,
we found that the layering of the intine was disorganized, uneven and absent in extreme
cases, resulting in a much wider gap between the nexine and the plasma membrane, and the
eventual collapse of developing microspores in *glcat14a/b* and *glcat14a/b/c* mutants
(**Figure 5-3A-3C; Appendix B: Supplemental Figure 5-2A-2F**). Consistent with the
observations made from TEM data, the collapse of the developing microspores was
similarly observed in *glcat14a/b* and *glcat14a/b/c* mutants relative to the WT when stained
with Toluidine Blue Dye (**Figure 5-3D-3F**). Surprisingly, in collapsed pollen grains, we
observed that the exine layer patterning was still intact despite the absence of cytoplasmic

contents in *glcat14a/b* and *glcat14a/b/c* mutants (**Appendix B: Supplemental Figure 5-2A-2F**).

Figure 5-3

Micrographs Showed Defects in Pollen Development in glcat14a/b and glcat14a/b/c

Mutants

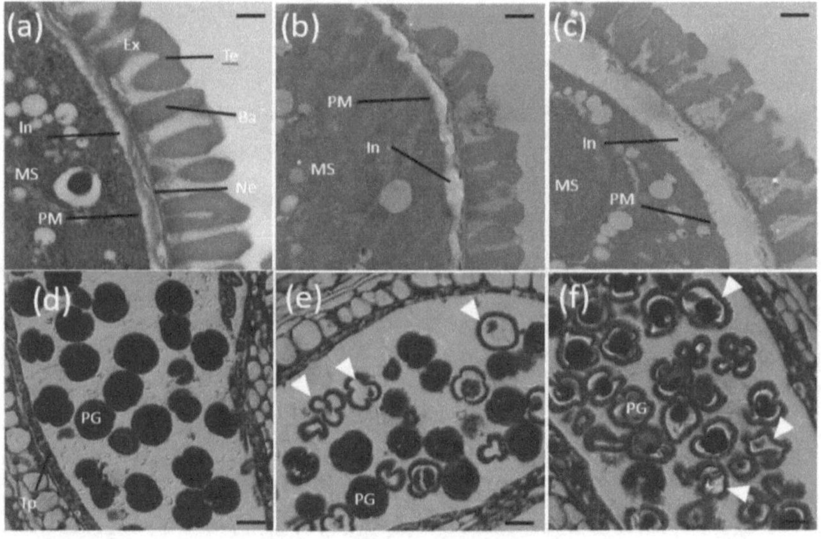

Note. TEM micrographs of mature pollen grains of WT (A), *glcat14a/b* (B) and *glcat14a/b/c* (C) mutants. Poorly arranged exine and abnormal development of the intine layers were observed in *glcat14a/b* and *glcat14a/b/c* pollen grains. Light micrographs of longitudinal sections of resin-embedded anthers of WT (D) *glcat14a/b* (E), and *glcat14a/b/c* (F) stained with toluidine blue. Notably, in contrast to WT, extensive cytoplasmic shrinkage and eventual collapse of pollen grains (arrowheads) were observed in *glcat14a/b* and *glcat14a/b/c* mutants. Ba, Bacula; Ex, exine; In, intine; MS, microspore; PG, pollen grain; PM, plasma membrane; Te, tectum; Tp, tapetal cell. Scale bars = 0.5µm in (A-C), 100 µm in (D-F).

***glcat14a/b* and *glcat14a/b/c* Mutants Showed Altered AGP Epitope Distribution**

AGPs role in microspore embryogenesis and pollen development is known to be essential to the reproductive success of several plant species (Costa et al., 2013, 2015; El-Tantawy et al., 2013; Pereira et al., 2014). To better understand the role of GLCATs in pollen development, anther cross sections were immunolabelled with AGP-specific monoclonal antibody, JIM13, and two pectin related monoclonal antibodies, JIM5 and JIM7 (that detects partially methylesterified and heavily methylesterified pectin respectively), to observe changes in the distribution of AGP and pectic polysaccharides and/or their methylesterification in Arabidopsis anther. At the pollen mother cell (PMC) stage, JIM13 strong labelling was observed in the wall that surrounds the microsporocytes that are dedifferentiating, and punctate signals were detected in the middle layer and the tapetal cells of WT. Also, bright JIM13 signals were seen in the cytoplasm of microsporocytes (**Figure 5-4A**). In *glcat14a/b* and *glcat14a/b/c* mutants, the JIM13 signal was not detected, except for the weak punctate signals around the endothecium (**Figure 5-4B and 4C**). Similarly, we found out that *glcat14a/b/c* mutants had lost the synchronous development between locules, which was present in the WT (**Compare asterisks in Figure 5-4C and Figure 5-4E**). Although we observed pollenless locules in *glcat14a/b/c* anthers, similar observations were made in the WT, albeit to a lower degree, while the intense labelling of JIM13 around the tapetum and endothecium were unaffected (**Figure 5-4D**). Surprisingly, JIM13 labelling in microspores at bicellular stages of *glcat14a/b* and *glcat14a/b/c* mutants were comparable to WT, except for increased collapse of pollen grains in *glcat14a/b/c* mutants (**Appendix**

B: Supplemental Figure 5-3A-3C). For JIM5, immunolabelling were comparatively similar in the anther locules of both WT and *glcat14a/b/c* mutants (**Appendix B: Supplemental Figure 5-4A and 4B**). For JIM7, anther from WT showed well developed locules with intense signal found around the tapetum and the endothecium, and also around the developing microspores at the uninucleate stage. Similar observations were made for *glcat14a/b* and *glcat14a/b/c* mutants, except for some aborted pollen grain lacking JIM7 immunolabelling around the pollen wall (**Appendix B: Supplemental Figure 5-4F**).

Figure 5-4

Distribution of JIM13-Epitope Labeling of AGPs of Arabidopsis Anthers at Different

Stages of Pollen Development in WT, glcat14a/b and glcat14a/b/c Mutants

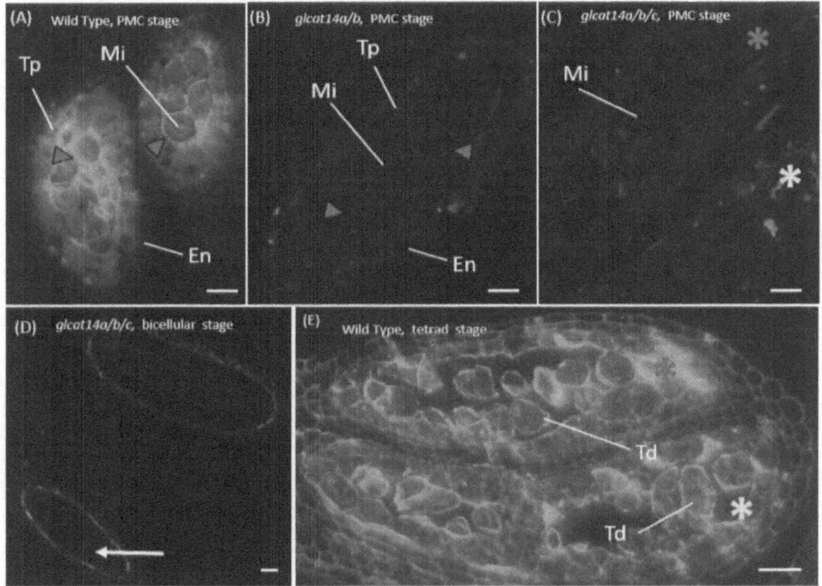

Note. Cross sections of resin-embedded anthers of WT at the pollen mother cell (PMC) stage, prior to meiosis, were labelled with JIM13 and subsequently with Alexa Fluor 488-labeled secondary antibody, showed intense labeling of the tapetal cells and the wall that surrounds the microsporocytes that are dedifferentiating (red arrows in A, B) (A); Contrast to WT, the intense labelling of JIM13 around the tapetal cells and the wall that surrounds the microsporocytes were absent in *glcat14a/b* (B) and *glcat14a/b/c* (C) mutants, except for the weak punctate signals observed around the endothecium. Pollenless locules were observed at bicellular stage in *glcat14a/b/c* anther (D), while intense labelling of JIM13 epitopes were observed in the WT tetrad stage (E). Arrow in d, indicates pollenless locules. Yellow and red asterisks in c, e, represents two locules that showed synchronous development in WT, tetrad stage (E) that was lacking in *glcat14a/b/c* anther in PMC stage (C). En, Endothecium; Mi, microsporocytes; PMC, pollen mother cell; Td, tetrad; Tp, tapetal cell. Bars = 100μm.

glcat14 mutants Showed Defects in Polytubey Block, Delayed Embryo Development and Reduced Fertility

In angiosperms, the production of viable offspring is reflected by its seed set, and it is central to continuity of plant life and performance. We observed that GLCAT14A and GLCAT14B are highly expressed in the micropyler endosperm region near the filiform apparatus, while GLCAT14C showed little or no expression (Winter et al., 2007) (**Appendix B: Supplemental Figure 5-5D**). Given the low seed set phenotype observed in *glcat14a/b/c* mutants (Zhang et al., 2020), we investigated further the underlying mechanisms responsible for the low seed set phenotype. We observed that the first five siliques of *glcat14a/b/c* mutants were significantly shorter than the WT (**Appendix B: Supplemental Figure 5-5A and 5B**). While it appears that the remainder of the siliques (excluding the first 5 siliques) of *glcat14a/b/c* mutants are comparable to WT, we found out that some seeds were aborted at different times in *glcat14a/b/c* mutants (**Figure 5-5A**), resulting in low seed set (**Figure 5-5D and 5E**). We next tested whether the reduced fertility was due to defective pollen. Although, the pollen tube lengths of WT were comparatively similar to *glcat14a/b* and *glcat14a/b/c* mutants (**Figure 5-5B**), we observed a reduction in LM2 immunolabelling at the pollen tube tip in *glcat14a/b* and *glcat14a/b/c* mutants relative to WT (**Figure 5-5C**), while JIM5 labelling of *glcat14a/b* and *glcat14a/b/c* were comparable to WT (**Appendix B: Supplemental Figure 5-5C**).

Figure 5-5

Seed Set and LM2 Immunolabeling of Pollen Tube Showed Defects in glcat14 Mutants

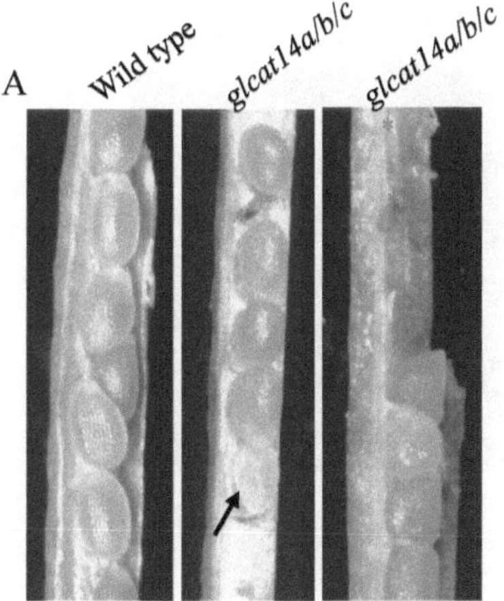

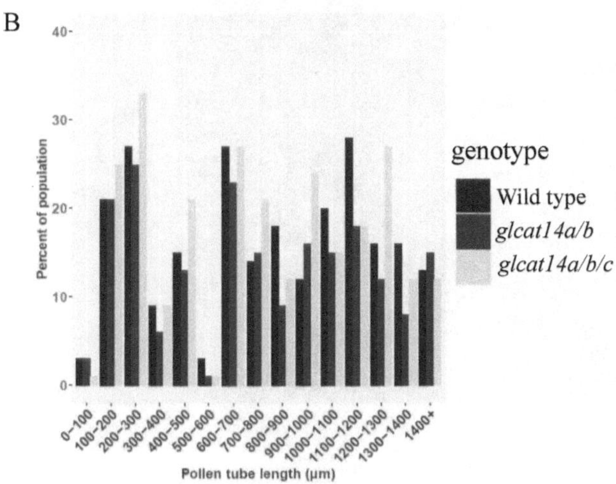

Figure 5-5: continued

LM2

C   Wild type   *glcat14a/b*   *glcat14a/b/c*

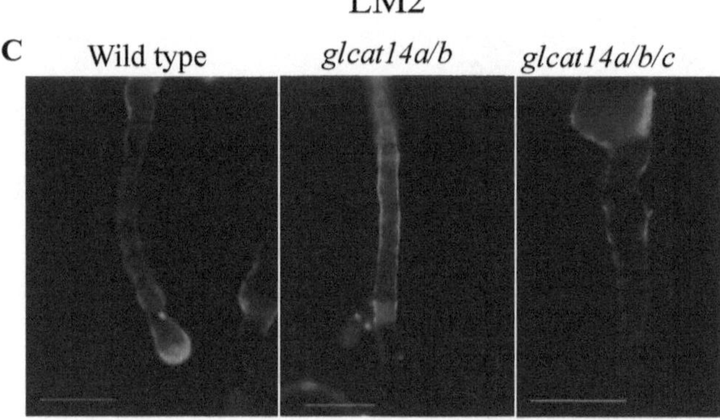

D   Wild type   *glcat14a/b*   *glcat14a/b/c*

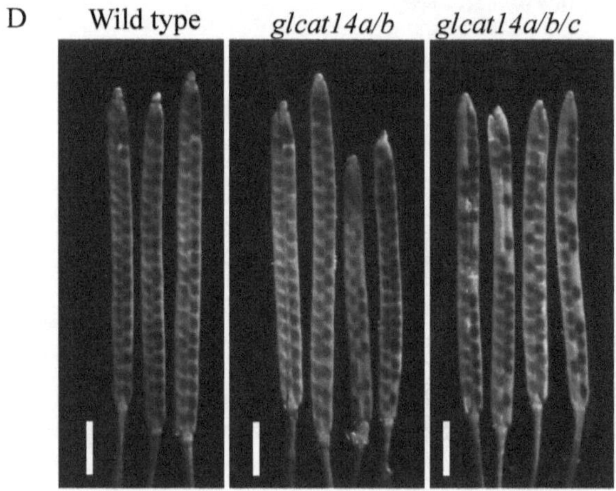

Figure 5-5: continued

E

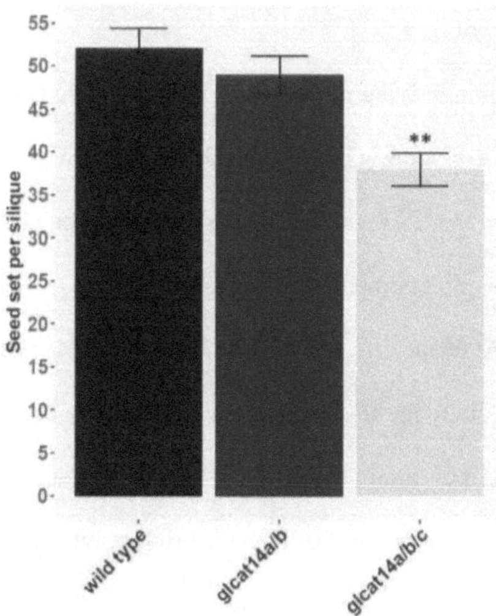

Note. (A) Representative silique opened to expose developing seeds of WT and *glcat14a/b/c* mutants showing defects in seed development. Red asterisks indicated aborted seeds that failed to develop; black arrow indicated seed that aborted at later stage. (B) Pollen tube lengths of WT and *glcat14* mutants (C) Images of LM2 immunolabelling of WT and *glcat14* mutants acquired using an epifluorescence microscope showed the absence of intense labelling especially at the tip region in *glcat14* mutants in contrast to WT. (D) Low seed set observed in *glcat14a/b/c* siliques. (E) Low seed set observed in *glcat14a/b/c* mutants compared to the WT (student's *t*-test, $P < 0.01$, $n \geq 20$ siliques per genotype). Scale bar in A= 2cm; C = 150μm, D = 1cm.

Interestingly, some of the *glcat14a/b/c* mutants with viable pollen grains germinated into pollen tubes and targeted the ovules most of the time (**Figure 5-6C**). However, to our surprise, in contrast to the WT (**Figure 5-6A and 6B**), we found out that some *glcat14a/b/c* ovules attracted excess number of pollen tubes (polytubey), implying defects in the polytubey block mechanism (**Figure 5-6D and 6E**), and a reciprocal cross showed that the polytubey phenotype was exacerbated when *glcat14a/b/c* ovule was used instead of the WT (**Figure 5-6E-6G**).

To better understand the underlying mechanisms mediating the polytubey phenotype, we immunolabeled Arabidopsis ovule using JIM5 and JIM13 monoclonal antibody. Previous work showed that FER-mediated presence of pectin at the filiform apparatus prevented the entry of supernumerary pollen tubes into the female gametophyte (Duan et al., 2020). Similarly, AGP specific antibodies (JIM8 and JIM13) specifically labelled the synergid cells, especially the filiform apparatus (Coimbra et al., 2007). We proceeded to use JIM5 and JIM13 to immunolabel the ovules of WT and *glcat14a/b/c* mutants, given the critical role of the synergid cells in ensuring polytubey block (Maruyama et al., 2015). While JIM5 immunolabelling of stage 12 flowers, which is the final stage before bud opens (Smyth et al., 1990) in WT were comparable to *glcat14a/b/c* mutants (**Figure 5-7A1 and 7A3**), we observed a reduction in JIM5 signal intensity at the micropyler ends in *glcat14a/b/c* mutants relative to the WT at stage 14 flowers, a stage where the long anther extends above the stigma (Smyth et al., 1990), which coincides with post-fertilization phase of Arabidopsis (**Figure 5-7A2 and 7A4**).

Figure 5-6

Aniline Blue Staining of WT, glcat14a/b/c Flowers, and their Reciprocal Crosses

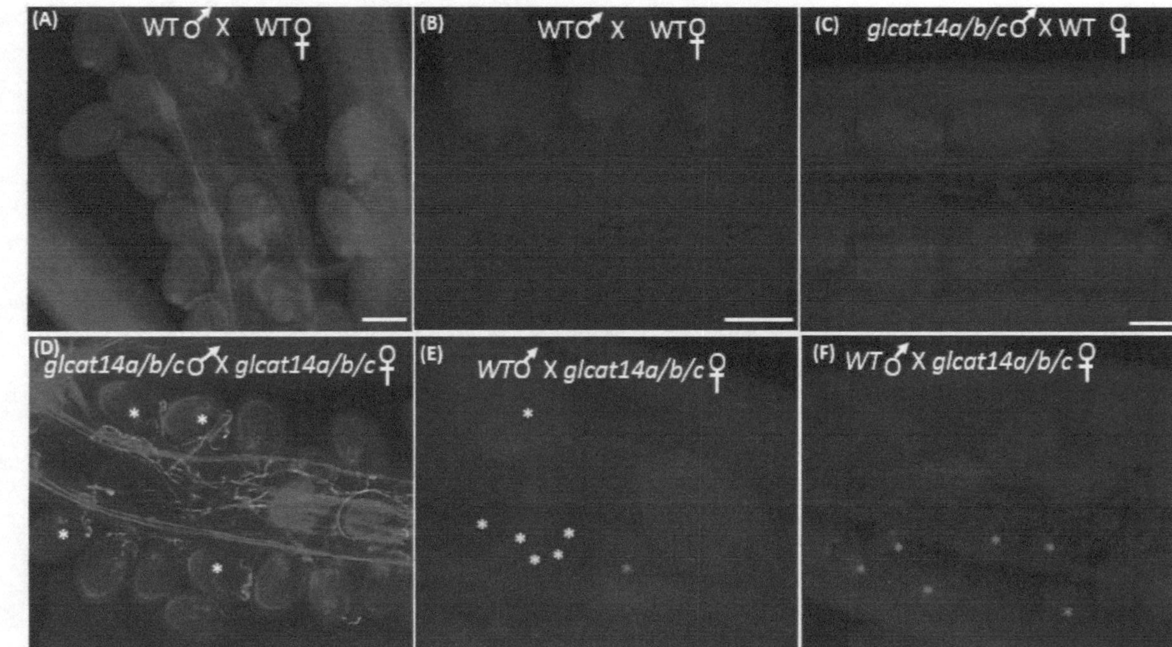

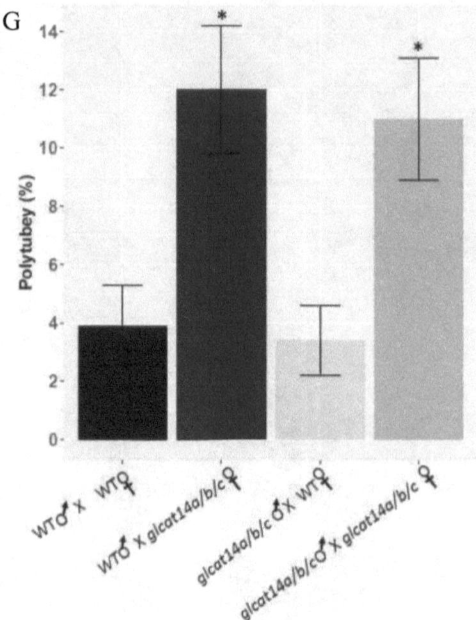

Note. A, B, D represents crosses between individuals of the same genotype as indicated; C, represents cross between *glcat14a/b/c* pollen and WT female; E, F, represents WT pollen and *glcat14a/b/c* female. No occurrence of polytubey were observed in A, B, C, as single pollen tube targeted individual ovary (indicated by arrow heads in panel B); In D and E, excess pollen tubes targeted single ovules, indicated by white asterisks in D, while up to five pollen tubes (indicated by yellow asterisks) branched from the transmission tract and targeted a single ovule with white asterisks; ovule in E, showed a seed with developmental defect that aborted at later stages (red asterisks), while some ovule failed to develop (red asterisks) as seen in F. G, represents the frequency of polytubey observed in WT, *glcat14a/b/c* mutants and their reciprocal crosses. WT × WT ($n = 250$), WT × *glcat14a/b/c* ($n = 325$), *glcat14a/b/c* × WT ($n = 396$), *glcat14a/b/c* X *glcat14a/b/c* ($n = 394$). Error bars denote mean ± SE, Asterisks indicates statistically significant, $P < 0.05$. Scale bar in A-F = 50 μm

Furthermore, JIM13 signal intensity were reduced in *glcat14a/b/c* mutants ovules at stage12 and stage14 flowers relative to the WT (**Figure 5-7B1-7B4**), and similar reduction in ruthenium red staining were observed at stage 14 and 15 (stigma extends above the long anther; Smyth *et al.,* 1990) of flower development in *glcat14a/b/c* mutant (**Figure 5-7C3-C6**). Given the different times seed abortion occur in *glcat14a/b/c* mutants, we examined the progress of seed development post fertilization in WT, *glcat14a/b* and *glcat14a/b/c* mutant ovules using vannilin, which has been previously used for the identification of fertilization defective mutants based on the pollen tube-dependent ovule enlargement morphology phenomenon, which generates a partial seed coat within the ovule without fertilization (Liu et al., 2020). Vanillin staining was used as a marker for seed coat development, which stains proanthocyanidins in the endothelium after fertilization (Debeauion et al., 2000). In contrast to the WT, the proportion of partially stained and unstained ovules were much more in *glcat14a/b/c* mutants (**Appendix B: Supplemental Figure 5-6A-6D**). Given the expression of GLCAT14A and GLCAT14B in micropyler endosperm, we asked if GLCATs play a role in embryo growth and we observed delays in embryo development in *glcat14a/b* and *glcat14a/b/c* mutants at the early globular, late globular, heart and bent stages relative to the WT (**Figure 5-8A-8D**).

Figure 5-7

Immunolabeling and Ruthenium Red Staining of WT and glcat14a/b/c Ovules

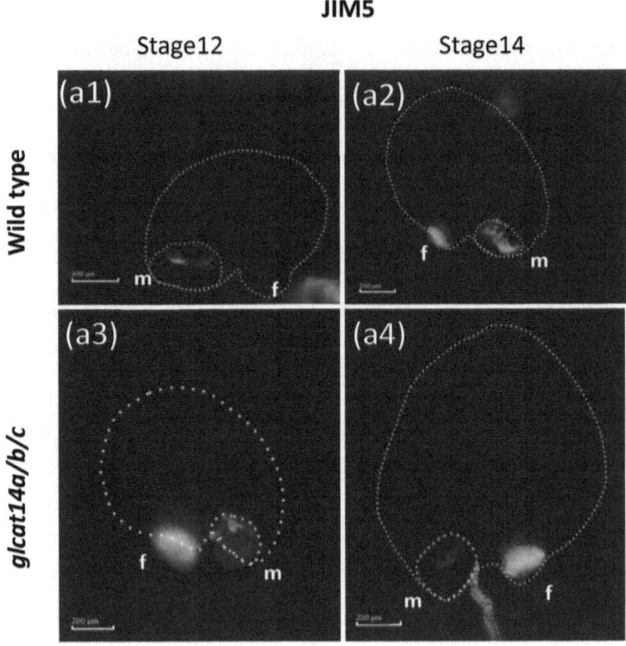

Figure 5-7: continued

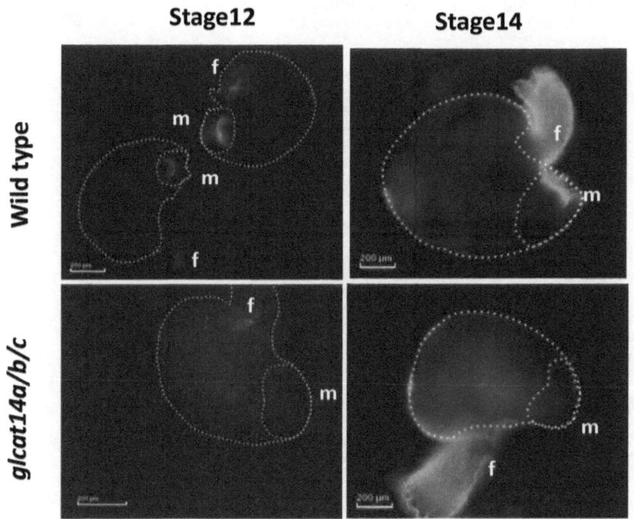

Figure 5-7: continued

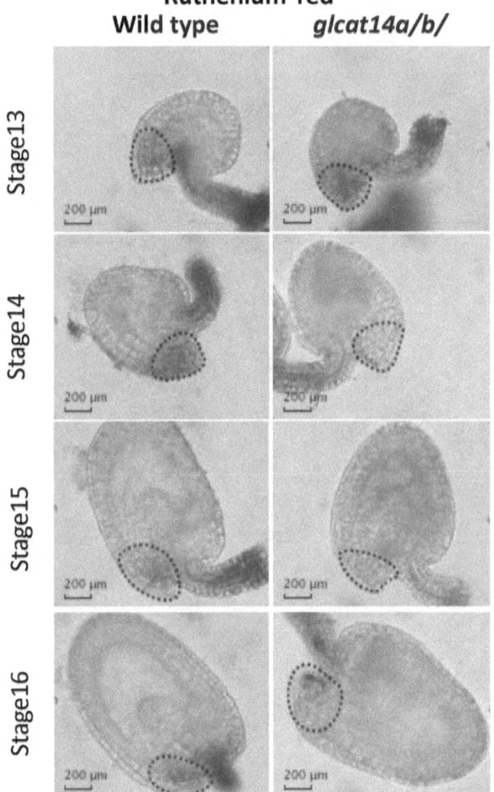

Note. In panel A, represents JIM5 immunolabeling of WT (a1, a2) and *glcat14a/b/c* ovules (a3, a4) at stage 12 (a1, a3) and stage 14 (a2, a4). JIM5 immunolabeling was comparatively similar in WT and *glcat14a/b/c* ovules at stage 12, but a reduction in JIM5 signal was observed in *glcat14a/b/c* relative to WT ovules at stage 14. In panel B, represents JIM13 immunolabeling of WT (b1, b2) and *glcat14a/b/c* ovules (b3, b4) at stage 12 (b1, b3) and stage 14 (a2, a4). In panel C, ruthenium red (RuR) detected pectin in ovules of WT and *glcat14a/b/c* mutants at the filiform apparatus of WT and *glcat14a/b/c* ovules, showed RuR reduction in *glcat14a/b/c* ovules at stages 14 (c3, c4) and 15 (c5, c6) relative to the WT. Floral stages followed that of previous publication (Smyth et al., 1990). m, micropylar region; f, funiculus

Figure 5-8

Embryo Development in WT, glcat14a/b and glcat14a/b/c Mutants

A Early globular

Wild type | *glcat14a/b* | *glcat14a/b/c*

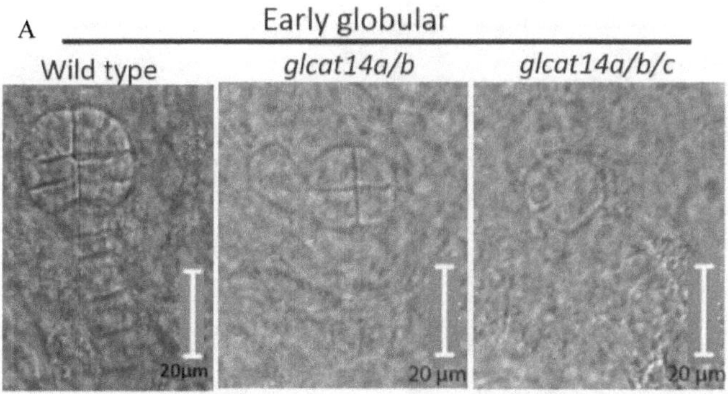

B Late globular

Wild type | *glcat14a/b* | *glcat14a/b/c*

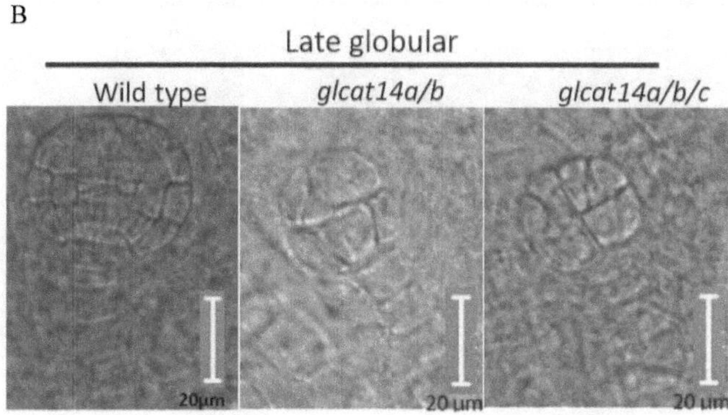

Figure 5-8: continued

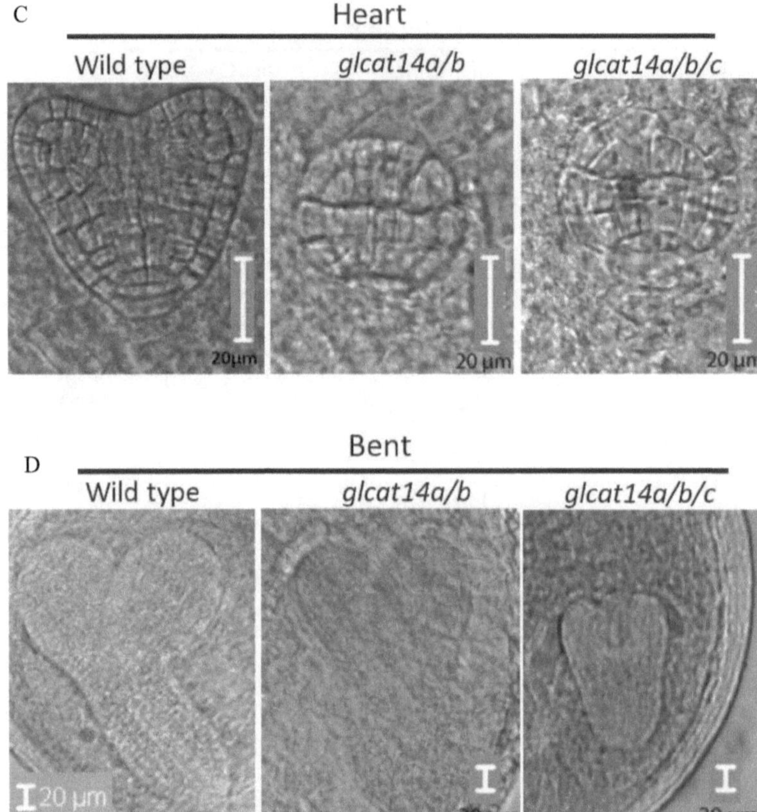

Note. Cleared developing seeds from the same silique of WT type, *glcat14a/b* and *glcat14a/b/c* mutants were examined and embryos from early globular (A), late globular (B), heart (C) and bent stages (D) of WT were observed to be different from *glcat14a/b* and *glcat14a/b/c*.

***glcat14* Mutants Showed Reduction in Glucuronic Acid Content With Altered Sugar Distribution**

Previous work showed that there are 11 GLCATs in GT14 family in Arabidopsis (Ajayi et al., 2020; Lopez-Hernandez et al., 2020; Ye et al., 2011) but with little insights into the contribution of members of these gene family in Arabidopsis flower development. Using monosaccharide composition analyses by HPAE-PAD and immunolabelling assay, we examined the glucuronic acid contents in flowers of *glcat14a/b* and *glcat14a/b/c* mutants relative to the WT. Results obtained in the monosaccharide composition analysis showed that relative to the WT, *glcat14a/b* and *glcat14a/b/c* mutants had 45% and 59% reduction in glucuronic acid contents, with a corresponding increase in arabinose and galactose contents (**Figure 5-9A**). Furthermore, changes in the composition of other sugars were observed in *glcat14a/b* and *glcat14a/b/c* mutants, especially rhamnose and galacturonic acid contents which are components of the RG-I backbone (**Figure 5-9A**). Interestingly, this observation was consistent with the results obtained when protein extracts from flower tissues were immunolabelled with LM2 and JIM13 monoclonal antibodies following SDS-PAGE. The binding of LM2, which recognizes a carbohydrate epitope containing β-linked GlcA, and JIM13 which recognizes (beta)GlcA1->3(alpha)GalA1->2Rha epitope (Knox et al., 1990), were found to be reduced in in *glcat14a/b* and *glcat14a/b/c* mutants relative to the WT (**Figure 5-9B**).

Figure 5-9

Monosaccharide Composition and Western Blot Analyses of WT and glcat14 Mutants

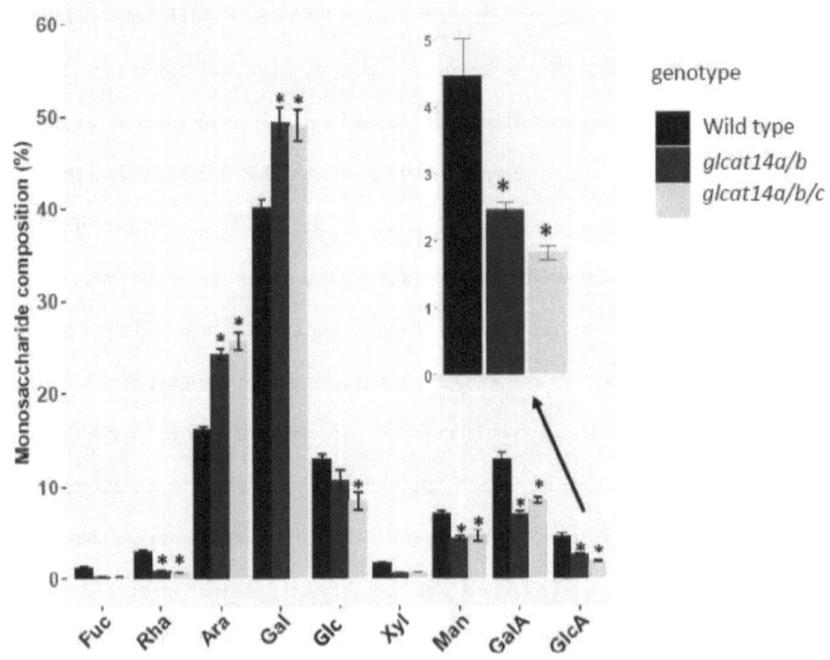

Figure 5-9: continued

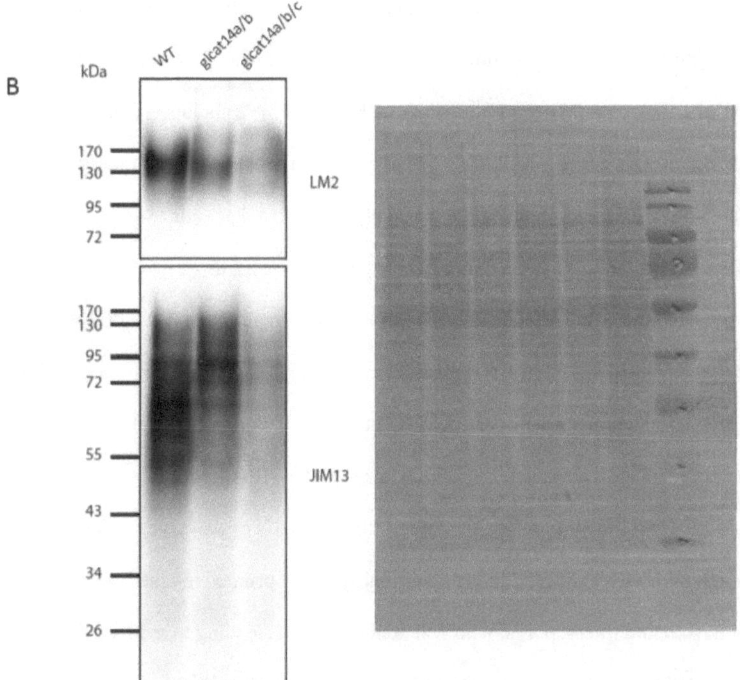

Note. (A) Sugar analysis of AGPs isolated from Arabidopsis flowers of WT and *glcat14* mutants showed reduction in all sugars except arabinose and galactose sugars. (B) Western blots (LM2, top; JIM13 bottom) and corresponding Ponceau-stained membrane (right) of flower protein samples of WT and *glcat14* mutants.

Discussion

AGPs are hyperglycosylated glycoproteins with little known about the contributions of its glycan moieties and protein backbones in sexual reproduction. Despite the potential occurrence of about 85 AGPs in Arabidopsis (Showalter et al., 2010), only a very few AGPs have been ascribed functions in both male and female gametophyte development, and extensively studied to date. For example, AGP18 has been implicated to be involved in female gametogenesis, as shown by RNAi experiments (Acosta-Garcı´a and Vielle-Calzada, 2004). Down-regulation of AGP6 and AGP11 by RNAi and artificial microRNA resulted in pollen grain abortion, inhibition of pollen tube growth, reduction of fertility (Levitin et al., 2008; Coimbra et al., 2009) and regulation of pollen germination via water uptake (Coimbra et al., 2010). Furthermore, KNS4/UPEX1 is a galactosyltransferase responsible for the synthesis of AG glycans and loss of function of KNS4/UPEX1 resulted in the abnormality of the primexine matrix laid on the surface of developing microspores, with concomitant effects leading to reduced fertility as indicated by shorter fruit lengths and lower seed set (Dobritsa et al., 2011; Li et al., 2017, Suzuki et al., 2017). In this work, we have extensively characterized three GLCATs, *GLCAT14A*, *GLCAT14B* and *GLCAT14C* genes involved in the glucuronidation of Type II AGs on AGPs and showed that they are critically important in Arabidopsis sexual reproduction. This study demonstrated that *GLCAT14A*, *GLCAT14B* and *GLCAT14C* genes are important in maintaining the reticulate architecture of pollen and pollen shape (**Figure 5-1A-1F and 1J**). Previously, defects in exine wall formation in *kns4/upex1* mutants were characterized and found to exhibit short baculae, small lacunae and over-developed tectum (Dobritsa et al., 2011, Li et al., 2017, Suzuki et al., 2017). This work showed severe alterations in the reticulate structure of mature pollen grains from *glcat14a/b* and *glcat14a/b/c* mutants with wide lacuna and the exposed bacula in *glcat14a/b* and *glcat14a/b/c* mutants (**Figure 5-1E and 1F**). Although, the reticulate architecture in *kns4/upex1* mutant appears to be slightly different from those observed in *glcat14a/b* and *glcat14a/b/c* mutants, our findings suggest that KNS4/UPEX1 and GLCAT14A, GLCAT14B and GLCAT14C likely function in the same genetic pathway but with different sites of action during pollen wall formation. We reasoned that this may be the case especially when galactosylation and glucuronidation of Type II AG polysaccharides are important events that are central to AGP glycan assembly and function. Furthermore, we found a significant proportion of misshaped (**Table 5-1**) and non-viable pollen in *glcat14a/b* and *glcat14a/b/c* mutants (**Figure 5-2**). AGPs have been implicated in pollen grain development as reported for a*gp6agp11* mutants (Coimbra et al., 2009). Therefore, we argued that the disruption of pollen specific AGPs and/or its glycan moieties such as glucuronic acid in this case, could impact pollen wall development.

Compared to the structural complexities involved in pollen exine assembly, the intine layer has a simpler structure, which is located between the cell membrane and the nexine of the exine (Ma et al., 2021). The main components of the intine are cellulose, hemicellulose, pectin, and structural protein, which are similar to those in the cell wall (Quilichini et al., 2015; Gigli et al., 2018). Previous work identified the role of *Arabidopsis FLA3* and *Brassica BcMF8* in pollen intine formation, while its involvement in cellulose deposition was speculated (Huang et al., 2008; Li et al., 2010; Lin et al., 2014). Our results from TEM analysis showed that the pollen intine layer formation was abnormal and uneven, and extensive shrinkage of the cytoplasm were observed, leading to the collapse of pollen grain as seen in *glcat14a/b* and *glcat14a/b/c* mutants (**Figure 5-3B, 3C and Appendix B: Supplemental Figure 5-2**). Given the reported connection between thickened intine layer and pollen stability (Marquez et al., 1997; Hesse, 2000; Shi et al., 2015), we infer that the collapse of the pollen grains might be due to the abnormal pollen intine layering in *glcat14a/b* and *glcat14a/b/c* mutants, evidenced by the absence of cytoplasmic contents in pollen grains of *glcat14a/*b and *glcat14a/b/c* mutants when stained with toluidine blue (**Figure 5-3E and 3F**). AGPs are structural proteins of the intine layer, and their silencing or deletion have been demonstrated to affect the structure of the intine (Jia et al., 2015). The presence of AGPs in the intine layer of *Quercus suber* (Costa et al., 2015) and the evidence of *Arabidopsis FLA3* and *Brassica BcMF8* function in intine development (Huang et al., 2008; Lin et al. 2014) led us to conclude that AG glucuronidation may be an important requirement for AGP function during intine development, and alteration in the AGP glycan structure may disrupt the pollen wall assembly process involving other cell wall polymers like pectin and cellulose, which are also components of the intine layer.

We further investigated pollen development by observing transverse sections of WT, *glcat14a/b* and *glcat14a/b/c* anther sections that were immunolabelled with AGP and pectin related antibodies (mAb) to understand the underlying events contributing to pollen abortion. Epitopes recognized by mAb JIM13 were intense around the wall surrounding the PMC of WT but absent in both *glcat14a/b* and *glcat14a/b/c* mutants (**Figure 5-4A-4C**). Interestingly, the intense labelling at the PMC stage during pollen development has been observed in previous studies involving microsporocytes immunolabelled with AGP specific (JIM8 and JIM13) mAb (Coimbra, 2007; Suzuki et al., 2018). But what was more surprising was that in contrast to WT, the synchronous development between locules within the same anther was lost in some *glcat14a/b/c* stamen (**Figure 5-4C, 4E**). Given that earlier works showed GlcA ability to bind to calcium (Lamport and Varnai, 2013; Lopez-Hernandez et al., 2020), and that calcium is important in anther development (Kong and Jia, 2004; Chen et al., 2008), we speculate that the calcium signaling process necessary for the proper coordination between locules may have been compromised and therefore warrants further investigation.

The fertilization process in angiosperms is structured in a way that a single pollen tube fertilizes an ovule through a polytubey block mechanism (Maruyama et al., 2013; Völz *et al.*, 2013). We observed that there was an excessive number of pollen tubes targeting *glcat14a/b/c* ovules (**Figure 5-6D and 6E**). Previous work identified an arabinogalactan protein, AGP4, named JAGGER, to be essential for persistent synergid degeneration and polytubey block, and JAGGER carbohydrate play a role in mediating polytubey block (Pereira et al., 2016). Maruyama et al. (2015) argued that after successful double fertilization, the persistent synergid degenerates following synergid-endosperm fusion, which causes rapid dilution of pre-secreted pollen tube attractant in the persistent synergid cell, and initiates the selective disorganization of the synergid nucleus during the endosperm proliferation to prevent polytubey. Similarly, Arabidopsis fail to block polytubey when the egg cell or central cell remain unfertilized, allowing the next pollen tube to recover the early fertilization failure (Beale et al., 2012; Kasahara et al., 2012; Maruyama et al., 2013). In addition, the synergid localized FERONIA through ovular pectins, have been demonstrated to be involved in polytubey block (Duan et al., 2020). The reduced JIM5 signal intensity at the micropyler ends of *glcat14a/b/c* ovules, the reduced immunolabelling of AGP specific JIM13 signals at the filiform apparatus close to the synergid cells, and the reduction in ruthenium red stainings in *glcat14a/b/c* mutant ovules (**Figure 5-7**) implies that the mechanistic processes involved in polytubey block may have been compromised in *glcat14a/b/c* mutants. This idea was further reinforced by the reduction in AGP-GlcA and pectin related sugars such as rhamnose and galacturonic acid that constitute pectin RG-I backbone, the reduction in LM2 immunolabelling, and reduction in JIM13 signal intensity in *glcat14a/b/c* mutants relative to WT (**Figure 5-9**). *Arabidopsis* AGP, AGP57C (At3g45230), was previously shown to covalently interact with pectin and hemicelluloses to form a continuous proteoglycan structure network proposed as ARABINOXYLAN PECTIN ARABINOGALACTAN PROTEIN1 (APAP1) (Tan et al., 2013). In light of this findings, we argued that the alteration in the AG glucuronidation impacted pectin formation and may have constitutively disrupted the persistent synergid inactivation process and polytubey block. This is important given that GLCAT14A, GLCAT14B and GLCAT14C are expressed in the micropyler region near the filiform apparatus (**Appendix B: Supplemental Figure 5-5D**).

The AGP-Ca^{2+} capacitor model suggests that AGPs can store and release Ca^{2+} in a pH-dependent manner (Lamport and Varnai, 2013; Lamport et al., 2018). The occurrence of calcium spikes that occur after gamete fusion, explains the importance of calcium in fertilization processes (Denninger et al., 2014; Hamamura et al., 2014). For successful fertilization, it is thought that cytoplasmic calcium oscillations are initiated in synergid cells after physical contact with the pollen tube tip, while an extended cytoplasmic calcium transiently occur in the egg cell (Denninger et al., 2014). In addition, an influx of calcium ions is thought to induce cell wall formation in the zygote (Kranz et al., 1995; Antoine et al., 2001). Also, the thin cell wall layer surrounding the central cell fuses with the persistent synergid cell immediately after fertilization, to avoid polytubey (Maruyama et al., 2015). Given the specific labelling of male gametes with AGP specific mAbs JIM8 (Coimbra, 2007) and JIM13 (**Appendix B: Supplemental Figure 5-3A**), coupled with the reduction in LM2 signal in the pollen tube tip of *glcat14a/b/c* mutants (**Figure 5-5C**), we reasoned that the reduction in AG glucuronidation may have altered the calcium dynamics necessary to ensure successful fertilization, and therefore act as an ancillary to the defect in polytubey block mechanism observed in *glcat14a/b/c* mutants. Understandably, the excessive amount of pollen tubes penetrating the female gametophyte may lead to polyspermy, and polyspermy in the central cell results in paternal and maternal genome imbalance in the endosperm that impacts normal seed development (Scott et al., 2008; Köhler et al., 2010). It is unclear to what extent genome imbalance may have contributed to the seed abortion and low seed set phenotypes observed in *glcat14a/b/c* mutants (**Figure 5-5A, 5D and 5E**), but we presume that the delay in embryo development observed in *glcat14a/b* and *glcat14a/b/c* mutants during seed development (**Figure 5-8**) may be due to the impairment in the fertilization of the central cell that gives rise to nutritious endosperm for the zygote (Köhler et al., 2010). Perhaps, calcium is necessary for accumulation and transport of amino acids and for carbohydrate metabolism (Ge et al., 2007) and thus, the alteration in calcium dynamics might have impacted normal seed development.

We were intrigued by the role of GLCAT14A, GLCAT14B and GLCAT14C in Arabidopsis sexual reproduction. How can these three genes impact pollen development and participate in polytubey block, while ensuring the normal seed development in angiosperms? Showalter et al. (2010) conducted a bioinformatic analysis identifying 85 AGPs in Arabidopsis, and their glycosylation is critical to their biological functions. We reasoned that GLCAT14A, GLCAT14B and GLCAT14C may be critically important in the glucuronidation of several AGPs involved in pollen development, such as AGP6 and AGP11 (Coimbra et al., 2009) and sexual fertilization processes, such as JAGGER (AGP4), which is required for persistent synergid degeneration and polytubey block (Pereira et al., 2016). We establish that the polytubey phenotype in *glcat14a/b/c* was not fully penetrant (**Figure 5-6G**), and similar concerns have been raised previously in *jagger* mutants (Pereira et al., 2016). Future work aimed at investigating AGP-GLCAT interactions with other molecules and pathways will bring us a step closer towards understanding the molecular mechanisms involved in sexual reproduction and reproductive development of angiosperms.

Chapter 6: Conclusions

AGPs are heavily glycosylated with type II arabinogalactan (AG) polysaccharides attached to hydroxyproline residues on their protein backbone. Despite the importance of type II AGs in plant growth processes, serving both structural and signaling functions (Lamport and Varnai, 2013), our understanding of the underlying role of these AG glycans/sugar residues in plant growth and development are poorly understood and far from complete. One such sugar residue is the GlcA of AGPs that are transferred onto AGP glycans by the actions of multiple β-GLCATs.

In other to better understand the AGP glucuronidation process, mediated by *GLCAT* genes belonging to GT14 family, my research focus was to functionally characterize three glucuronosyltransferase genes involved in AGP biosynthesis and investigate their roles in Arabidopsis growth and development. Prior to this work, only three members of this GT14 family of enzymes, GLCAT14A (AT5G39990), GLCAT14B (AT5G15050) and GLCAT14C (AT2G37585) were functionally characterized using an *in-vitro* assay and demonstrated to have GLCAT activity by adding glucuronic acid (GlcA) to β-1,3 and β-1,6-linked galactans of AG glycans (Dilokpimol and Geshi, 2014; Knoch et al., 2014). Previous work using nuclear magnetic resonance and molecular dynamics simulations demonstrated that the terminal carboxyl groups of GlcA residues can act as potential intramolecular Ca^{2+}-binding sites (Lamport and Varnai, 2013). Apart from the fact that GlcA residues are the only acidic sugars in the AG chain, their importance in the binding and release of extracellular calcium have been demonstrated (Lamport and Varnai, 2013; Lopez-Hernandez et al., 2020). Intuitively,

GLCATs may play an essential role in global Ca^{2+} signaling processes in plants and could be a potential target for improving AGP-based products and increasing biomass for biofuel production from bioenergy crops.

This book presents important findings related to the biochemical and physiological roles of *GLCAT14A*, *GLCAT14B* and *GLCAT14C* genes using genetic mutant analysis. Given that other GT14 family members exist, with little information about their evolutionary relationships, the first project was to use bioinformatics tools to mine plant genomes for evolutionary footprints of functional importance among putative GLCATs by using the *Arabidopsis thaliana* GLCAT domain as queries. The specific research goal was to utilize phylogenetic analysis, physical mapping of genes, motif identification, synteny and gene expression analyses to gain additional insight into the evolution, sequence diversification, functional similarity/divergence, and tissue-specific transcriptional profiling among putative *GLCAT* gene family members in 14 plant genomes. A total of 161 putative *GLCAT* genes were found distributed across 14 plant genomes and were characterized by a widely conserved GLCAT domain which is the catalytic region with less conserved N terminal and C terminal ends following protein sequence alignments (**Figure 2-1**). The assessment of the phylogenetic relationships among putative GLCAT protein sequences using both maximum likelihood approach and bayesian inferences identified distinct phylogenetic clades, with each having species representatives in each subclade (**Figure 2-2A**). Apart from the widely conserved exon-intron architecture of the putative *GLCATs* (**Figure 2-3**), RNA-seq and microarray data analyses of the putative *GLCAT* genes revealed gene expression signatures that are likely important in plant growth processes (**Figure 2-6**). One such unique gene expression signatures for *GLCAT14A, GLCAT14B* and *GLCAT14C* genes were found in the Arabidopsis seed coat. These three GLCAT genes have different expression patterns in the seed coat, and in order to elucidate their potential roles in seed coat development, single and higher order *glcat* mutants (**Figure 3-1B**) were generated and their phenotypes compared to the WT. To this end, two *GLCAT* genes, *GLCAT14A* and *GLCAT14C*, involved in the biosynthesis of type II AGs were discovered to function in mucilage polysaccharide matrix organization in *Arabidopsis thaliana*.

One model system that is gaining increasing recognition and significance for the study of cell wall polysaccharide interactions is the Arabidopsis seed coat epidermis (SCE), also referred to as Mucilage Secretary Cells (MSC) (Griffiths and North, 2017). The SCE is an excellent model system for understanding the genetic basis of cell wall biosynthesis, secretion, assembly and modification (Voiniciuc et al., 2015) because large amounts of cell wall polysaccharides can be extracted with ease and analyzed in a short timeframe. I used this approach to better understand the role of GLCATs in plant cell wall biosynthesis and wall formation. To this end, *in-silico* analysis of *GLCAT14A, GLCAT14B* and *GLCAT14C* showed that *GLCAT14A* and *GLCAT14C* genes were much more expressed in the seed coat, compared to *GLCAT14B* that showed little or no expression (**Figure 3-1A**), In addition, seed mucilage defects were observed in single *glcat14a-1* but were further exacerbated in the *glcat14a-1glcat14c-1* mutants, characterized by a significant loss of adherent mucilage (**Figure 3-3**) and significant alterations in cellulose ray formation (**Figure 3-5**) and seed coat morphology

(**Figure 3-8**). Monosaccharide composition analysis showed significant alterations in the sugar amounts of *glcat14a-1 glcat14c-1* mutants relative to WT in the adherent and non-adherent seed mucilage (**Table 3-1**). Also, reduction in total mucilage content was observed in *glcat14a-1glcat14c-1* mutants relative to WT (**Figure 3-7**), while *glcat14a-1glcat14c-1* mutants showed defects in pectin formation, calcium content and the degree of pectin methyl-esterification (DM) (**Figure 3-6),** as well as reductions in crystalline cellulose content (**Figure 3-5**) and seed size (**Figure 3-9**). SOS5 has been extensively characterized and implicated in seed mucilage biosynthesis and glycosylated SOS5 has been demonstrated to interact with FEI1/FEI2 (Basu et al., 2016). It would be necessary to investigate further, whether GLCAT14A and GLCAT14C act in a linear, non-additive pathway with GALT2, GALT5, SOS5, FEI1, and FEI2. Also, it is worth unraveling whether GLCAT14A and GLCAT14C act on some or all of AGPs involved in mucilage formation and also whether other GLCATs are involved in stabilizing seed mucilage organization. Also, it is unclear what role the *GLCAT14A* and *GLCAT14C* genes play in stabilizing the covalent attachment of rhamnosyl residues of pectin RG-I backbone to AGPs in APAP1 (Tan et al., 2013) and this presents an enigma worth unraveling.

Plant cell walls are highly recalcitrant and research efforts have been directed at deconstructing plant cell wall polysaccharides to facilitate the accessibility of cellulose rich cell wall to cellulases. Plant cell wall polysaccharides are hydrolyzed using acid, base or enzymes but progress has been slowed down due to the recalcitrant nature of plant cell walls. The loss of adherent mucilage in *glcat14a-1glcat14c-1* mutants in this work offers the potential of genetically manipulating plant cell wall materials and brings us a step closer towards the identification of key genes that ease the extraction of mucilage materials in its native form for the design of glycoconjugates.

The roles of GLCAT14A, GLCAT14B and GLCAT14C in the vegetative tissue development of Arabidopsis were delineated using biochemical and physiological approaches. Previous work showed that GLCAT14A was localized to the Golgi (Knoch et al., 2013), and similar observations were made for GLCAT14B and GLCAT14C when they were transiently expressed in *Nicotiana benthamiana* (**Figure 4-6**). Sugar analysis of AGP extracts from Arabidopsis stem, leaf and siliques showed a consistent reduction in glucuronic acid in *glcat14* mutants relative to WT, with concomitant effects resulting in tissue specific alterations, especially in arabinose and galactose sugars (**Figure 4-4**). In addition, mutant phenotypes such as defects in trichome branching and plant height were much more severe in *glcat14a/b* and *glcat14a/b/c* (**Figure 4-4E and Figure 4-7A**) and findings were consistent with earlier studies (Zhang et al., 2020). Despite the presence of 11 GLCATs in Arabidopsis, only five genes (*GLCAT14A, GLCAT14B* and *GLCAT14C, GLCAT14D* and *GLCAT14E*) have been characterized and found critically important in multiple developmental processes in Arabidopsis (Knoch et al., 2020; Zhang et al., 2020; Lopez-Hernandez et al., 2020). An important area of research is to investigate the biochemical and physiological functions of other GLCATs in plant growth and developmental processes. Also, the mechanism of action of the remaining GLCATs is worth investigating. For example, do all GLCATs in GT14 family require same or different AG glycan substrates in order to glucuronidate AGPs and what is the site of action for these GLCATs? What is the nature of redundancy among GT14 family members in AGP glucuronidation in vegetative tissue development? Given that GlcA

undergoes different modifications (Tryfona et al., 2013), which *GLCATs* are methylated,

bound to calcium or crosslinked with other heteropolysaccharides and how does this

sugar modifications impact biological functions? Notably, AG from *glcat14a/b/d* triple

mutants had lower calcium binding capacity *in vitro* than AG from WT (Lopez-

Hernandez et al., 2020), but the underlying connections between calcium bound AGP and

plant growth still remains an enigma worth unraveling.

The role of AG glucuronidation in plant growth and development has been

revealed through this work. But given the role of AGP in calcium binding (Lopez-

Hernandez et al., 2020) and the important role that calcium plays in plant growth, future

work aimed at genetically engineering GLCATs to potentially enhance calcium's role in

improving plant biomass for bioenergy production is an important step worth

considering.

The biological roles of GLCAT14A, GLCAT14B and GLCAT14C are not limited

to mucilage formation or vegetative development but are also involved in reproductive

development and sexual fertilization in Arabidopsis. Double fertilization is a unique

reproductive system shared by angiosperms for the successful development of viable

offspring. AGPs are complex hyperglycosylated cell wall proteins that function in plant

sexual reproduction (Coimbra et al., 2007). Although *AtAGP6* and *AtAGP11* were

implicated in pollen development (Coimbra et al., 2009) and AGP4 has been reported to

be involved in polytubey block (Pereira et al., 2016), it was unclear what roles

GLCAT14A, GLCAT14B and GLCAT14C play in reproductive development and sexual fertilization processes. Previous work on *glcat14aglcat14bglcat14c* mutants showed that they have defective pollen grains, shorter siliques and reduced seed set (Zhang et al., 2020). However, the underlying reasons were unknown. I extended this work to further investigate the molecular function of *GLCAT14A*, *GLCAT14B* and *GLCAT14C* in Arabidopsis sexual development. Based on findings from this work, *GLCAT14A*, *GLCAT14B* and *GLCAT14C* were redundant, as defects in Arabidopsis sexual reproduction were only observed in *glcat14a/b* and *glcat14a/b/c* mutants relative to other single and higher order mutant combinations. Specifically, this work showed that GLCAT14A, GLCAT14B and GLCAT14C function in pollen development, polytubey block, and normal embryo development in Arabidopsis. The disruption of the reticulate structure of the exine wall (**Figure 5-1**), abnormal development of the intine layer (**Figure 5-3A-3C**), and collapse of pollen grains (**Figure 5-3D-3F**) were observed in *glcat14a/b* and *glcat14a/b/c* mutants when compared to WT. The proportion of non-viable pollen in *glcat14a/b* and *glcat14a/b/c* mutants were higher than WT (**Table 5-1**) and this finding was consistent with Alexander and Ruthenium red staining (**Figure 5-2**). Given that AGPs play important roles in microspore embryogenesis and pollen development (Costa et al., 2013, 2015; El-Tantawy et al., 2013; Pereira et al., 2014), the absence of JIM13 signals in the anther cross sections of *glcat14a/b* and *glcat14a/b/c* mutants relative to WT at the PMC stage is reflective of the important roles AGP play at the early stages of pollen development (**Figure 5-4B and 4C**). Apart from the reduced LM2 immunolabelling at the pollen tube tip observed in both *glcat14a/b* and *glcat14a/b/c* mutants (**Figure 5-5C**), the attraction of excess numbers of pollen tubes (polytubey) was apparent in *glcat14a/b/c* ovules, implying that the polytubey block mechanism was compromised (**Figure 5-6**). FERONIA (FER)-mediated presence of pectin at the filiform apparatus showed the importance of pectin in preventing the entry of supernumerary pollen tubes into the female gametophyte (Duan et al., 2020). While we found that JIM5 and JIM13 signals were reduced at the filiform apparatus in *glcat14a/b/c* ovules, how the reductions in AGP and pectin compromised polytubey block mechanisms remains unknown and as such, warrants further investigation. For example, how does the reduced AGP and pectin epitopes at the filiform apparatus mediates persistent synergid degeneration and polytubey block? Could AGPs work synergistically with pectins to mediate polytubey block? What is the nature of their interaction in the polytubey block pathway? AGP4 (JAGGER) has been reported to be essential for persistent synergid degeneration and polytubey block (Pereira et al., 2016), but is AG glucuronidation also critical to AGP4 function? Are other AGPs, in addition to AGP4, also glucuronidated by GLCAT14A, GLCAT14B and GLCAT14C to maximize reproductive success? Apart from GLCAT14A, GLCAT14B and GLCAT14C, are there other GLCATs in GT14 family involved in Arabidopsis sexual reproduction? What is the role of calcium bound AGPs in preventing polytubey and maximizing reproductive success and what is the exact mechanism of action? Understandably, glucuronic acid can either be methylated (Tryfona et al., 2012), calcium bound (Lamport and Varnai, 2013), crosslinked with hetero-polysaccharides (Tan et al., 2013) or unsubstituted. It would be interesting to know how tissue-specific modifications of glucuronic acid sugar affect the reproductive physiology of plant species. For example, previous work in *Torenia fournieri* demonstrated that a disaccharide, methyl-glucuronosyl galactose from AGPs, is essential to induce competence in Torenia pollen tubes, making them capable of recognizing the embryo sac molecular attractants, LUREs (Mizukami et al., 2016). Notably, my research work was done in Arabidopsis, it would also be interesting to investigate whether certain GlcA modifications confer unique biological functions regarding plant reproductive development and sexual fertilization in angiosperms. Interestingly, GLCAT14A, GLCAT14B and GLCAT14C were also observed to be critically important in establishing normal embryo development (**Figure 5-8A-8D**) but further work is needed to understand the mechanism of action regarding GLCATs role in embryo development in Arabidopsis.

Considering the important roles GLCAT14A, GLCAT14B and GLCAT14C play in Arabidopsis sexual reproduction, how can these three genes impact pollen development and participate in polytubey block, while ensuring normal seed development in angiosperms? Showalter et al. (2010) conducted a bioinformatic analysis identifying 85 AGPs in Arabidopsis, and their glycosylation is most likely critical to their biological functions given that carbohydrate forms the interactive molecular surface of AGPs. Are GLCAT14A, GLCAT14B and GLCAT14C the only GLCATs that are involved in the glucuronidation of some or all AGPs involved in Arabidopsis sexual reproduction? Interestingly, previous work showed that *glcat14a/b/e* were sterile (Lopez-Hernandez et al., 2020), but the underlying mechanism responsible for the defect remains to be elucidated. Currently, there are speculations that calcium bound AGPs contributes to the global calcium signaling pathways of green plants (Lamport et al., 2014) but what calcium gates/channels are involved in mediating the intracellular flow of AGP bound calcium and how they coordinate with intracellular calcium stores to modulate calcium signaling pathways remains an enigma worth unravelling.

As sexual fertilization success is critical to continuity of plant life, the insight gained from this work regarding the role of AG glucuronidation in sexual reproduction could serve as a foundation for investigations towards elucidating the underlying mechanisms involved in maximizing reproductive development, sexual fertilization success and embryo development in other angiosperms. Future work using next-generation sequencing (RNA-seq) would help identify other molecules and pathways that are mediated by AGP-GLCATs and will bring us a step closer towards understanding the molecular mechanisms involved in ensuring normal vegetative growth and reproductive development of angiosperms.

Finally, this work demonstrated the important roles that AGP-GLCATs play in plant growth and development. Due to the contribution of fossil fuels to greenhouse gas emissions, research efforts have been directed at increasing the use of bio-renewable materials as sustainable energy sources for biofuel production. Given that angiosperms life starts from the seed and that plant life/biomass is critical to bioenergy production, the roles that GLCATs play in vegetative and reproductive development, sexual fertilization, embryo development and mucilage formation underscores the fact that GLCATs could serve as potential gene targets for the sustainable development of plant biomass and for engineering plants with useful traits for the field and research laboratory.

References

Abramoff, M. D., Magalhaes, P. J. & Ram, S. J. (2004). Image processing with ImageJ. Biophotonics Int. 11, 36–42.

Acosta-Garciéa, G. & Vielle-Calzada, J. P. (2004) A classical arabinogalactan protein is essential for the initiation of female gametogenesis in Arabidopsis. Plant Cell. 16, 2614–2628.

Ajayi, O. O. & Showalter, A. M. (2020). Systems identification and characterization of β-glucuronosyltransferase genes involved in arabinogalactan-protein biosynthesis in plant genomes. Sci. Rep. 10, 20562.

Albersheim, P., Darvill, A., Roberts, K., Sederoff, R. & Staehelin, A. (2010). Plant cell walls from chemistry to biology. Garland Science, New York.

Alexander, M. P. (1969) Differential staining of aborted and non-aborted pollen. Stain Technol. 41, 117– 122.

Alonso, J. M., Stepanova, A. N., Leisse, T. J., Kim, C. J., Chen, H. et al. (2003). Genome-wide insertional mutagenesis of Arabidopsis thaliana. Science. 301, 653–657.

Amborella Genome Project. (2013). The Amborella genome and the evolution of flowering plants. Science. 342, 1241089.

Anderson, C. T. (2016). 'We be jammin': an update on pectin biosynthesis, trafficking and dynamics. J. Exp. Bot. 67, 495–502.

Anderson, C. T., Carroll, A., Akhmetova, L. & Somerville, C. (2010). Real-time imaging of cellulose reorientation during cell wall expansion in Arabidopsis roots. Plant Physiol. 152, 787–796.

Anderson, C. T. & Kieber, J. J. (2020). Dynamic construction, perception, and remodeling of plant cell walls. Annu. Rev. Plant Biol. 71, 39-69.

Antoine, A. F., Dumas, C., Faure, J. E., Feijó, J. A. & Rougier, M. (2001). Egg activation in flowering plants. Sex. Plant Reprod. 14, 21–26.

Ariizumi, T. & Toriyama, K. (2011). Genetic regulation of sporopollenin synthesis and pollen exine development. Annu. Rev. Plant. Biol. 62, 1–24.

Arioli, T., Peng, L., Betzner, A. S., Burn, J., Wittke, W. et al. (1998). Molecular analysis of cellulose biosynthesis in Arabidopsis. Science. 279, 717–720.

Arsovski, A. A., Haughn, G. W. & Western, T. L. (2010). Seed coat mucilage cells of Arabidopsis thaliana as a model for plant cell wall research. Plant Signaling and Behaviour. 5, 796-801.

Asif, M. H., Trivedi, P. K., Misra, P. & Nath, P. (2009). Prolyl-4-hydroxylase (AtP4H1) mediates and mimics low oxygen response in *Arabidopsis thaliana*. Funct. Integr. Genomics. 9, 525–535.

Atmodjo, M. A., Hao, Z. & Mohnen, D. (2013). Evolving views of pectin biosynthesis. Annu. Rev. Plant Biol. 64, 747–779.

Atmodjo, M. A., Sakuragi, Y., Zhu, X., Burrell, A. J., Mohanty, S. S. et al. (2011). Galacturonosyltransferase (GAUT)1 and GAUT7 are the core of a plant cell wall pectin biosynthetic homogalacturonan: galacturonosyltransferase complex. Proc. Natl. Acad. Sci. U. S. A. 108, 20225–20230.

Bailey, T., Boden, M., Buske, F., Frith, M., Grant, C. E. et al. (2009). MEME SUITE. Tools for motif discovery and searching. Nucleic Acids Res. 37, W202–W208.

Baldwin, T. C., McCann, M. C. & Roberts, K. (1993). A novel hydroxyproline deficient arabinogalactan protein secreted by suspension-cultured cells of Daucus carota: purification and partial characterization. Plant Physiol. 103, 115–123.

Bascom, C. S., Hepler, P. K. & Bezanilla, M. (2018). Interplay between ions, the cytoskeleton, and cell wall properties during tip growth. Plant Physiol. 176, 28–40.

Basu, D., Liang, Y., Liu, X., Himmeldirk, K., Faik, A. et al. (2013). Functional identification of a hydroxyproline-O-galactosyltransferase specific for arabinogalactan protein biosynthesis in Arabidopsis. J. Biol. Chem. 288, 10132–10143.

Basu, D., Tian, L., DeBrosse, T., Poirier, E., Emch, K. et al. (2016). Glycosylation of a fasciclin-like arabinogalactan-protein (SOS5) mediates root growth and seed mucilage adherence via a cell wall receptor-like kinase (FEI1/FEI2) pathway in Arabidopsis. PLoS ONE. 11, e0145092.

Basu, D., Tian, L., Wang, W., Bobbs, S., Herock, H. et al. (2015). A small multigene hydroxyproline-O-galactosyltransferase family functions in arabinogalactan-protein glycosylation, growth and development in Arabidopsis. B. M. C. Plant Biol. 15:, 295.

Batoko, H., Zheng, H. Q., Hawes, C. & Moore, I. (2000). A rab1 GTPase is required for transport between the endoplasmic reticulum and golgi apparatus and for normal Golgi movement in plants. Plant Cell. 11, 2201-18.

Bayer, M., Nawy, T., Giglione, C., Galli, M., Meinnel, T. et al. (2009). Paternal control of embryonic patterning in Arabidopsis thaliana. Science. 10, 1485-8.

Beale, K. M., Leydon, A. R. & Johnson, M. A. (2012). Gamete fusion is required to block multiple pollen tubes from entering an Arabidopsis ovule. Curr. Biol. 22, 1090–1094.

Bisceglia, N. G., Engelsdorf, T. & Hamann, T. (2020). Plant cell wall integrity maintenance in model plants and crop species-relevant cell wall components and underlying guiding principles. Cell. Mol. Life Sci. 77, 2049-2077.

Bisceglia, N. G., Savatin, D.V., Cervone, F., Engelsdorf, T. & Lorenzo, G.D. (2018). Loss of the *Arabidopsis* protein kinases ANPs affects root cell wall composition and triggers the cell wall damage syndrome. Front. Plant Sci. 8, 2234.

Blake, A. W., McCartney, L., Flint, J. E., Bolam, D. N., Boraston, A. B., et al. (2006). Understanding the biological rationale for the diversity of cellulose-directed carbohydrate-binding modules in prokaryotic enzymes. J. Biol. Chem. 281, 29321–29329.

Blumenkrantz, N. & Asboe-Hansen, G. (1973). New method for quantitative determination of uronic acids. Anal Biochem. 54, 484-489.

Borner, G. H. H., Lilley, K. S., Stevens, T. J. & Dupree, P. (2003). Identification of glycosylphosphatidylinositol-anchored proteins in Arabidopsis: A proteomic and genomic analysis. Plant Physiol. 132, 568–577.

Braccini, I. & Perez, S. (2001). Molecular basis of Ca^{2+}-induced gelation in alginates and pectins: the egg-box model revisited. Biomacromolecules. 2, 1089–1096.

Bradford, M. M. (1976). Rapid and sensitive method for quantitation of microgram quantities of protein utilizing principle of protein-dye binding. Anal. Biochem. 72, 248-254.

Brady, J. D., Sadler, I. H. & Fry, S. C. (1998). Pulcherosine, an oxidatively coupled trimer of tyrosine in plant cell walls: its role in cross-link formation. Phytochemistry. 47, 349–353.

Camirand, A. & Maclachlan, G. (1986). Biosynthesis of the fucose-containing xyloglucan nonasaccharide by pea microsomal membranes. Plant Physiol. 82, 379–383.

Cannon, M. C., Terneus, K., Hall, Q., Tan, L., Wang, Y. et al. (2008). Self-assembly of the plant cell wall requires an extensin scaffold. Proc. Natl. Acad. Sci. U. S. A. 105, 2226–2231.

Cantarel, B. L., Coutinho, P. M., Rancurel, C., Bernard, T., Lombard, V. et al. (2009). The Carbohydrate-Active EnZymes database (CAZy): an expert resource for glycogenomics. Nucleic Acids Res. 37, D233-238.

Carroll, A., Mansoori, N., Li, S., Lei, L., Vernhettes, S. et al. (2012). Complexes with mixed primary and secondary cellulose synthases are functional in Arabidopsis plants. Plant Physiol. 160, 726–737.

Carroll, S. B., Grenier, J. K. & Weatherbee, S. D. (2005). Evolution at two levels: On genes and form. PLoS Biol. 3, 1159–1166.

Chen, S. H., Liao, J. P., Luo, M. Z. & Kirchoff, B. K. (2008). Calcium distribution and function during anther development of Torenia fournieri (Linderniaceae). Annales Botanici Fennici. 45, 195–203.

Chen, X., Vega-Sanchez, M. E., Verhertbruggen, Y., Chiniquy, D., Canlas, P. E. et al. (2013). Inactivation of OsIRX10 leads to decreased xylan content in rice culm cell walls and improved biomass saccharification. Mol. Plant. 6, 570–573.

Cheung, A. Y., Wang, H. & Wu, H. M. (1995). A floral transmitting tissue-specific glycoprotein attracts pollen tubes and stimulates their growth. Cell. 82, 383–393.

Cheung, A. Y. & Wu, H. M. (1999). Arabinogalactan proteins in plant sexual reproduction. Protoplasma. 208, 87–98.

Chiniquy, D., Sharma, V., Schultink, A., Baidoo, E. E., Rautengarten, C. et al. (2012). XAX1 from glycosyltransferase family 61 mediates xylosyltransfer to rice xylan. Proc. Natl. Acad. Sci. U. S. A. 109, 17117–17122.

Citovsky, V., Lee, L. Y., Vyas, S., Glick, E., Chen, M. H. et al. (2006). Subcellular localization of interacting proteins by bimolecular fluorescence complementation in planta. J Mol Biol. 362, 1120–1131.

Clausen, M. H., Willats, W. G. & Knox, J. P. (2003). Synthetic methyl hexagalacturonate hapten inhibitors of anti-homogalacturonan monoclonal antibodies LM7, JIM5 and JIM7. Carbohydr. Res. 338, 1797–1800.

Coimbra, S., Almeida, J., Junqueira, V., Costa, M. L. & Pereira, L. G. (2007). Arabinogalactan proteins as molecular markers in Arabidopsis thaliana sexual reproduction. J. Exp.Bot. 58, 15-16.

Coimbra, S., Costa, M., Jones, B., Mendes, M. A. & Pereira, L. G. (2009). Pollen grain development is compromised in Arabidopsis agp6 agp11 null mutants. J. Exp. Bot. 60, 3133–3142.

Coimbra, S., Costa, M., Mendes, M. A., Pereira, A. M., Pinto, J. et al. (2010). Early germination of Arabidopsis pollen in a double null mutant for the arabinogalactan protein genes AGP6 and AGP11. Sex. Plant Reprod. 3, 199-205.

Corns, C. M. & Ludman, C. J. (1987). Some observations on the nature of the calcium cresolphthalein complexone reaction and its relevance to the clinical laboratory. Ann. Clin. Biochem. 24, 345–51.

Cosgrove, D. J. (2014). Plant cell growth and elongation. Chichester, UK: John Wiley & Sons, Ltd.

Costa, M., Nobre, M. S., Becker, J. D., Masiero, S., Amorim, M. I. et al. (2013). Expression-based and co-localization detection of arabinogalactan protein 6 and arabinogalactan protein 11 interactors in Arabidopsis pollen and pollen tubes. B. M. C. Plant Biol. 13, 7.

Costa, M. L., Sobral, R., Costa, M. M. R., Amorim, M. I. & Coimbra, S. (2015). Evaluation of the presence of arabinogalactan proteins and pectins during *Quercus suber* male gametogenesis. Ann. Bot. 11, 81–92.

Coyne, J. A. & Hoekstra, H. E. (2007). The locus of evolution: Evo devo and the genetics of adaptation. Evol. Int. J. Organ. Evol. 61(5), 995–1016.

Culbertson, A. T., Ehrlich, J. J., Choe, J. Y., Honzatko, R. B. & Zabotina, O. A. (2018). Structure of xyloglucan xylosyltransferase 1 reveals simple steric rules that define biological patterns of xyloglucan polymers. Proc. Natl. Acad. Sci. U. S. A. 115, 6064–6069.

Dardelle, F., Lehner, A., Ramdani, Y., Bardor, M., Lerouge, P. et al. (2010). Biochemical and immunocytological characterizations of Arabidopsis pollen tube cell wall. Plant Physiol. 4, 1563-76.

Darzentas, N. (2010). Circoletto: Visualizing sequence similarity with Circos. Bioinformatics 26, 2620–2621.

Debeaujon, I., Léon-Kloosterziel, K. M. & Koornneef, M. (2000). Influence of the testa on seed dormancy, germination, and longevity in Arabidopsis. Plant Physiol. 122, 403–413.

Denninger, P., Bleckmann, A., Lausser, A., Vogler, F., Ott, T. et al. (2014). Male–female communication triggers calcium signatures during fertilization in Arabidopsis. Nat. Commun. 5, 4645.

Desprez, T., Juraniec, M., Crowell, E. F., Jouy, H., Pochylova, Z., et al. (2007). Organization of cellulose synthase complexes involved in primary cell wall synthesis in *Arabidopsis thaliana*. PNAS. 104, 15572–77.

Dhugga, K. S., Barreiro, R., Whitten, B., Stecca, K., Hazebroek, J. et al. (2004). Guar seed beta-mannan synthase is a member of the cellulose synthase super gene family. Science. 303, 363–366.

Dilokpimol, A. & Geshi, N. (2014). *Arabidopsis thaliana* glucuronosyltransferase in family GT14. Plant Signal. Behav. 9, e28891.

Dilokpimol, A., Poulsen, C. P., Vereb, G., Kaneko, S., Schulz, A. et al. (2014). Galactosyltransferases from *Arabidopsis thaliana* in the biosynthesis of type II arabinogalactan: molecular interaction enhances enzyme activity. B. M. C. Plant Biol. 14, 90.

Doblin, M. S., Pettolino, F. A., Wilson, S. M., Campbell, R., Burton, R. A. et al. (2009). A barley cellulose synthase-like CSLH gene mediates (1,3;1,4)-β-D-glucan synthesis in transgenic Arabidopsis. Proc. Natl. Acad. Sci. U. S. A. 106, 5996–6001.

Dobritsa, A. A., Geanconteri, A., Shrestha, J., Carlson, A., Kooyers, N. et al. (2011). A large-scale genetic screen in Arabidopsis to identify genes involved in pollen exine production. Plant Physiol. 157, 947–970.

Drummond, A. J. & Rambaut, A. (2007). BEAST: Bayesian evolutionary analysis sampling trees. B. M. C. Evol. Biol. 7, 214.

Du, J., Mansfield, S. D. & Groover, A. T. (2009). The Populus homeobox gene
ARBORKNOX2 regulates cell differentiation during secondary growth. Plant J.
60, 1000-1014.

Duan, Q., Kita, D., Johnson, E. A., Aggarwal, M., Gates, L. et al. (2014) Reactive oxygen
species mediate pollen tube rupture to release sperm for fertilization in
Arabidopsis. Nat. Commun. 5, 3129.

Duan, Q., Liu, M. J., Kita, D., Jordan, S. S., Yeh, F. L. J. et al. (2020). FERONIA
controls pectin- and nitric oxide-mediated male–female interaction.
Nature. 579, 561–566.

Dubois, M., Gilled, K., Hamilton, J. K., Rebers, P. A. & Smith, F. (1956). Colorimetric
method for the determination of sugars and related substances. Anal. Chem. 28,
350–356.

Ebert, B., Birdseye, D., Liwanag, A. J. M., Laursen, T., Rennie, E. A. et al. (2018). The
three members of the Arabidopsis glycosyltransferase family 92 are functional
beta-1,4-galactan synthases. Plant Cell Physiol. 59, 2624–2636.

Egelund, J., Petersen, B. L., Motawia, M. S., Damager, I., Faik, A. et al. (2006).
Arabidopsis thaliana RGXT1 and RGXT2 encode Golgi-localized (1,3)-alpha-D-
xylosyltransferases involved in the synthesis of pectic rhamnogalacturonan-II.
Plant Cell. 18, 2593–2607.

Ellis, M., Egelund, J., Schultz, C. J., & Bacic, A. (2010). Arabinogalactan-proteins: key
regulators at the cell surface? Plant Physiol. 153, 403–419.

El-Tantawy, A. A., Solís, M. T., Costa, M. L., Coimbra, S., Risueño, M. C. et al. (2013). Arabinogalactan protein profiles and distribution patterns during microspore embryogenesis and pollen development in *Brassica napus*. Plant Reprod. 26, 231–243.

Estevez, J. M., Kieliszewski, M. J., Khitrov, N. & Somerville, C. (2006). Characterization of synthetic hydroxyproline-rich proteoglycans with arabinogalactan protein and extensin motifs in Arabidopsis. Plant Physiol. 142, 458–470.

Faik, A. (2010). Xylan biosynthesis: news from the grass. Plant Physiol. 153, 396–402.

Farrokhi, N., Burton, R. A., Brownfield, L., Hrmova, M. et al. (2006). Plant cell wall biosynthesis: genetic, biochemical and functional genomics approaches to the identification of key genes. Plant Biotechnol. J. 4, 145–167.

Finkelstein, R. R., Gampala, S. S. & Rock, C. D. (2002). Abscisic acid signaling in seeds and seedlings. Plant Cell. 14, 15-45.

Fragkostefanakis, S., Dandachi, F. & Kalaitzis, P. (2012). Expression of arabinogalactan proteins during tomato fruit ripening and in response to mechanical wounding, hypoxia and anoxia. Plant Physiol. Biochem. 52, 112–118.

Fry, S. C. (2000). The Growing Plant Cell Wall: Chemical and Metabolic Analysis. (Caldwell, McCartney NJ: The Blackburn Press). pg. 34-43.

Ganguly, A., Zhu, C., Chen, W. & Dixit, R. (2020). FRA1 kinesin modulates the lateral stability of cortical microtubules through cellulose synthase-microtubule uncoupling proteins. Plant Cell. 32, 2508–2524.

Gao, F., Chen, C., Arab, D. A., Du, Z., He, Y. et al. (2019). EasyCodeML: A visual tool for analysis of selection using CodeML. Ecol. Evolut. 9, 3891–3898.

Gaspar, Y., Johnson, K. L., McKenna, J. A., Bacic, A. & Schultz, C. J. (2001). The complex structures of arabinogalactan-proteins and the journey towards understanding function. Plant Molecular Biology. 47, 161-176.

Gasteiger, E., Hoogland, C., Gattiker, A., Duvaud, S., Wilkins, M. R. et al. (2005). Protein identification and analysis tools on the ExPASy Server. In The Proteomics Protocols Handbook (ed. Walker, J. M.) 571–607.

Ge, L. L., Tian, H. Q. & Russell, S. D. (2007). Calcium function and distribution during fertilization in angiosperms. Am J Bot. 6, 1046-60.

Geshi, N., Johansen, J. N., Dilokpimol, A., Rolland, A., Belcram, K. et al. (2013). A galactosyltransferase acting on arabinogalactan protein glycans is essential for embryo development in Arabidopsis. Plant J. 76, 128–137.

Gille, S., Hänsel, U., Ziemann, M. & Pauly, M. (2009). Identification of plant cell wall mutants by means of a forward chemical genetic approach using hydrolases. Proc. Natl. Acad. Sci. U. S. A. 106, 14699–14704.

Gille, S., Sharma, V., Baidoo, E. E. K., Keasling, J. D., Scheller, H. V. et al. (2013). Arabinosylation of a yariv-precipitable cell wall polymer impacts plant growth as exemplified by the arabidopsis glycosyltransferase mutant ray1. Mol. Plant. 6, 1369–1372.

Gimeno-Ferrer, F., Pastor-Cantizano, N., Bernat-Silvestre, C., Selvi-Martı´nez, P., Vera-Sirera, F. et al. (2017). a2-COP is involved in early secretory traffic in Arabidopsis and is required for plant growth. J. Exp. Bot. 68, 391–401.

Gómez-Maqueo, X. & Gamboa-deBuen, A. (2016). The dynamics of plant cell wall in muro modifications and its physiological implications on seed germination. InTech.

Goodstein, D. M., Shu, S., Howson, R, Neupane, R., Hayes, R. D. et al. (2012). Phytozome: A comparative platform for green plant genomics. Nucleic Acids Res. 40, D1178–D1186.

Götting, C., Müller, S., Schöttler, M., Schön, S., Prante, C. et al. (2004). Analysis of the DXD motifs in human Xylosyltransferase I required for enzyme activity. J. Biol. Chem. 41, 42566–42573.

Griffiths, J. S., Crepeau, M. J., Ralet, M. C., Seifert, G. J. & North, H. M. (2016). Dissecting Seed Mucilage Adherence Mediated by FEI2 and SOS5. Front. Plant Sci. 7, 1073.

Griffiths, J. S. & North, H. M. (2017). Sticking to cellulose: exploiting Arabidopsis seed coat mucilage to understand cellulose biosynthesis and cell wall polysaccharide interactions. New Phytologist. 214, 959–966.

Griffiths, J. S., Tsai, A. Y. L., Xue, H., Voiniciuc, C., Sola, K. et al. (2014). SALT-OVERLY SENSITIVE5 mediates Arabidopsis seed coat mucilage adherence and organization through pectins. Plant Physiol. 165, 991–1004.

Gou, M., Ran, X., Martin, D. W. & Liu, C. J. (2018). The scaffold proteins of lingin biosynthetic cytochrome P450 enzymes. Nat. Plants. 4, 299–310.

Gu, Y., Kaplinsky, N., Bringmann, M., Cobb, A., Carroll, A. et al. (2010). Identification of a cellulose synthase-associated protein required for cellulose biosynthesis. Proc. Natl. Acad. Sci. U. S. A. 107, 12866–12871.

Guindon, S., Rodrigo, A. G., Dyer, K. A. & Huelsenbeck, J. P. (2004). Modeling the site-specific variation of selection patterns along lineages. Proc. Natl. Acad. Sci. U. S. A. 101, 12957–12962.

Guo, Y., Xiong, L., Song, C., Gong, D., Halfter, U. et al. (2002). A calcium sensor and its interacting protein kinase are global regulators of abscisic acid signaling in Arabidopsis. Developmental Cell. 3, 233-244.

Hamamura, Y., Nishimaki, M., Takeuchi, H., Geitmann, A., Kurihara, D. et al. (2014). Live imaging of calcium spikes during double fertilization in Arabidopsis. Nat. Commun. 5, 4722.

Hamamura, Y., Saito, C., Awai, C., Kurihara, D., Miyawaki, A. et al. (2011) Live-cell imaging reveals the dynamics of two sperm cells during double fertilization in Arabidopsis thaliana. Curr. Biol. 21, 497–502.

Hamann, T. (2015). The plant cell wall integrity maintenance mechanism–A case study of a cell wall plasma membrane signaling network. Phytochemistry. 112, 100–109.

Hamann, T. & Denness, L. (2011). Cell wall integrity maintenance in plants: Lessons to be learned from yeast? Plant Signaling & Behavior. 6, 1706–1709.

Han, K. K. & Martinage, A. (1992). Possible relationship between coding recognition amino acid sequence motif or residue(s) and post-translational chemical modification of proteins, Int. J. Biochem. 24, 1349–1363.

Harholt, J., Suttangkakul, A. & Vibe Scheller, H. (2010). Biosynthesis of Pectin. Plant Physiol. 153, 384–395.

Harpaz-Saad, S., McFarlane, H. E., Xu, S., Divi, U. K., Forward, B., et al. (2008). Cellulose synthesis via the FEI2 RLK/SOS5 pathway and CELLULOSE SYNTHASE 5 is required for the structure of seed coat mucilage in Arabidopsis. Plant Journal. 68, 941–953.

Harpaz-Saad, S., Western, T. L. & Kieber, J. J. (2012). The FEI2-SOS5 pathway and CELLULOSE SYNTHASE 5 are required for cellulose biosynthesis in the Arabidopsis seed coat and affect pectin mucilage structure. Plant Signal Behav. 7, 285–288.

Harris, M. R. & Hofmann, H. A. (2015). Seeing is believing: Dynamic evolution of gene families. Proc. Nat. Acad. Sci. 112(5), 1252–1253.

Haughn, G. W., & Western, T. L. (2012). Arabidopsis seed coat mucilage is a specialized cell wall that can be used as a model for genetic analysis of plant cell wall structure and function. Front. Plant Sci. 3, 64.

Held, M. A., Tan, L., Kamyab, A., Hare, M., et al. (2004). Di-isodityrosine is the intermolecular cross-link of isodityrosine-rich extensin analogs cross-linked in vitro. J. Biol. Chem. 279, 55474–5548210.

Hepler, P. K. (2005). Calcium: a central regulator of plant growth and development. Plant Cell. 17, 2142-2155.

Hesse, M. (2000). Pollen wall stratification and pollination. Plant Syst. Evol. 222, 1–17.

Hill, J. L. Jr., Hammudi, M. B. & Tien, M. (2014). The Arabidopsis cellulose synthase complex: a proposed hexamer of CESA trimers in an equimolar stoichiometry. Plant Cell. 26, 4834–42.

Hoffmann, N., King, S., Samuels, A. L. & McFarlane, H. E. (2021). Subcellular coordination of plant cell wall synthesis. Dev Cell. 56(7), :933-948.

Hou, W. C., Chang, W. H., & Jiang, C. M. (1999). Qualitative distinction of carboxyl group distribution in pectins with ruthenium red. Bot. Bull. Acad. Sin. 40, 115–119.

Hu, B., Jin, J., Guo, A. Y., Zhang, H., Luo, J., et al. (2015). GSDS 2.0: An upgraded gene feature visualization server. Bioinformatics. 31, 1296–1297.

Hu, R., Li, J., Wang, X., Zhao, X., Yang, X., et al. (2016). Xylan synthesized by Irregular Xylem 14 (IRX14) maintains the structure of seed coat mucilage in Arabidopsis. J Exp Bot. 5, 1243-57.

Huang, L., Cao, J. S., Zhang, A. H. & Ye, Y. Q. (2008). Characterization of a putative pollen-specific arabinogalactan protein gene, *BcMF8*, from *Brassica campestris* ssp. *chinensis*. Mol Biol Rep. 35, 631–639.

Huang, J., Ju, Y., Wang, X. & Zhang, S. (2015). A onestep rectification of sperm cell targeting ensures the success of double fertilization. J. Integr. Plant Biol. 57, 496–503.

Huang, Y., Wang, Y., Tan, L., Sun, L., Petrosino, J. et al. (2016). Nanospherical arabinogalactan proteins are a key component of the high-strength adhesive secreted by English ivy. PNAS. 23, E3193-E3202.

Humphrey, T. V., Bonetta, D. T. & Goring, D. R. (2007). Sentinels at the wall: cell wall receptors and sensors. New Phytol. 176, 7–21.

Immerzeel, P., Eppink, M. M., De Vries, S. C. et al. (2006). Carrot arabinogalactan proteins are interlinked with pectins. Plant Physiol. 128, 18–28.

Ishii, T. (1997). Structure and functions of feruloylated polysaccharides. Plant Sci. 127, 111–127.

Ishii, T. & Matsunaga, T. (1996). Isolation and characterization of a boron-rhamnogalacturonan-II complex from cell walls of sugar beet pulp. Carbohydr. Res. 284, 1–9.

Iwai, H., Masaoka, N., Ishii, T. & Satoh, S. (2002). A pectin glucuronyltransferase gene is essential for intercellular attachment in the plant meristem. Proc. Natl. Acad. Sci. U. S. A. 99, 16319–16324.

Jia, Q. S., Zhu, J., Xu, X. F., Lou, Y., Zhang, Z. L. et al. (2015). *Arabidopsis* AT-hook protein TEK positively regμlates the expression of arabinogalactan proteins for nexine formation. Mol. Plant. 8, 251–260.

Jiang, N., Wiemels, R.E., Soya, A., Whitley, R., Held, M., Faik, A. (2016). Composition, assembly, and trafficking of a wheat xylan synthase complex. Plant Physiol. 70, 1999-2023.

Jiao, S., Hazebroek, J. P., Chamberlin, M. A., Perkins, M., Sandhu, A. S. et al. (2019). Chitinase-like1 plays a role in stalk tensile strength in maize. Plant Physiol. 181, 1127–1147.

Johnson, K. L., Jones, B. J., Bacic, A. & Schultz, C. J. (2003). The fasciclin-like arabinogalactan proteins of arabidopsis. A multigene family of putative cell adhesion molecules. Plant Physiol. 133, 1911–25.

Josè-Estanyol, M. & Puigdomènech, P. (2000). Plant cell wall glycoproteins and their genes. Plant Physiology and Biochemistry. 38, 97-108.

Kang, B. H., Nielsen, E., Preuss, M. L., Mastronarde, D. & Staehelin, L. A. (2011). Electron tomography of RabA4b- and PI-4Kb1-labeled trans Golgi network compartments in Arabidopsis. Traffic. 12, 313–329.

Kang, X., Kirui, A., Dickwella Widanage, M. C., Mentink-Vigier, F., Cosgrove, D. J. et al. (2019). Lignin-polysaccharide interactions in plant secondary cell walls revealed by solid-state NMR. Nat. Commun. 10, 347.

Kasahara, R. D., Portereiko, M. F., Sandaklie-Nikolova, L., Rabiger, D. S. & Drews, G. N. (2005). MYB98 is required for pollen tube guidance and synergid cell differentiation in Arabidopsis. Plant Cell. 17, 2981–2992.

Keskiaho, K., Hieta, R., Sormunen, R. & Myllyharju, J. (2007). Chlamydomonas reinhardtii has multiple prolyl 4-hydroxylases, one of which is essential for proper cell wall assembly. Plant Cell. 19, 256–269.

Kieber, J. J. & Polko, J. (2019). The regulation of cellulose biosynthesis in plants. Plant Cell. 31, 282–296.

Kieliszewski, M. J. (2001). The latest hype on Hyp-O-glycosylation codes. Phytochemistry. 57, 319–323.

Kieliszewski, M. J. & Lamport, D. T. A. (1994). Extensin: repetitive motifs, functional sites, post-translational codes, and phylogeny, Plant J. 5, 157–172.

Kim, S. J., Chandrasekar, B., Rea, A. C., Danhof, L., Zemelis-Durfee, S. et al. (2020). The synthesis of xyloglucan, an abundant plant cell wall polysaccharide, requires CSLC function. Proc. Natl. Acad. Sci. U. S. A. 117, 20316–20324.

Kim, S. J., Held, M. A., Zemelis, S., Wilkerson, C. & Brandizzi, F. (2015). CGR2 and CGR3 have critical overlapping roles in pectin methylesterification and plant growth in *Arabidopsis thaliana*. Plant J. 82, 208–220.

Kitazawa, K., Tryfona, T., Yoshimi, Y., Hayashi, Y., Kawauchi, S. et al. (2013). β-galactosyl Yariv reagent binds to the β-1,3-galactan of arabinogalactan proteins. Plant Physiol. 161, 1117–1126.

Klavons, J. A & Bennett, R. D. (1986). Determination of Methanol Using Alcohol Oxidase and Its Application to Methyl-Ester Content of Pectins. J. Agri. Food Chem. 34, 597-599.

Knoch, E., Dilokpimol, A. & Geshi, N. (2014). Arabinogalactan proteins: focus on carbohydrate active enzymes. Front. Plant Sci. 5, 198.

Knoch, E., Dilokpimol, A., Tryfona, T., Poulsen, C. P., Xiong, G. et al. (2013). A β-glucuronosyltransferase from *Arabidopsis thaliana* involved in biosynthesis of type II arabinogalactan has a role in cell elongation during seedling growth. Plant J. 76, 1016–1029.

Knox, J. P., Linstead, P. J., Cooper, J. P. C. & Roberts, K. (1991). Developmentally
regulated epitopes of cell surface arabinogalactan proteins and their relation to
root tissue pattern formation. Plant J. 1, 317–326.

Knox, J. P., Linstead, P. J., King, J., Cooper, C., & Roberts, K. (1990). Pectin
esterification is spatially regulated both within cell walls and between developing
tissues of root apices. Planta. 181, 512-521.

Köhler, C., Scheid, O. M. & Erilova, A. (2010). The impact of the triploid block on the
origin and evolution of polyploid plants. Trends Genet. 26, 142–148.

Kong, H. Y. & Jia, G. X. (2004). Calcium distribution during pollen development of
Larix principis-rupprechtii. Acta. Botanica. Sinica. 46, 69–76.

Kong, Y., Zhou, G., Yin, Y., Xu, Y., Pattathil, S. et al. (2011). Molecular analysis of a
family of Arabidopsis genes related to galacturonosyltransferases. Plant Physiol.
155, 1791–1805.

Koski, M. K., Hieta, R., Böllner, C., Kivirikko, K. I., Myllyharju, J. et al. (2007). The
active site of an algal prolyl 4-hydroxylase has a large structural plasticity. The
active site of an algal prolyl 4-hydroxylase has a large structural plasticity. J. Biol.
Chem. 282, 37112–37123.

Kranz, E., Wiegen, P. & Lörz, H. (1995). Early cytological events after induction of cell
division in egg cells and zygote development following in vitro fertilization with
angiosperm gametes. Plant J. 8, 9–23.

Krupkova, E., Immerzeel, P., Pauly, M. & Schmulling, T. (2007). The TUMOROUS
SHOOT DEVELOPMENT2 gene of Arabidopsis encoding a putative

methyltransferase is required for cell adhesion and co-ordinated plant development. Plant J. 50, 735–750.

Lairson, L. L., Henrissat, B., Davies, G. J. & Withers, S. G. (2008). Glycosyltransferases: Structures, functions, and mechanisms. Annu. Rev. Biochem. 77, 521–555.

Lamport, D. T. A. (1963). Oxygen fixation into hydroxyproline of plant cell wall protein. J. Biol. Chem. 238, 1438–1440.

Lamport, D. T. A. (2013). Preparation of arabinogalactan glycoproteins from plant tissue. BIO-Protoc. 3, 1-3.

Lamport, D. T. A., Kieliszewski, M. J. & Showalter, A. M. (2006). Salt stress upregulates periplasmic arabinogalactan proteins: Using salt stress to analyse AGP function. New Phytol. 169, 479–492.

Lamport, D. T. A., Tan, L., Held, M. A. & Kieliszewski, M. J. (2018). Pollen tube growth and guidance: Occam's razor sharpened on a molecular arabinogalactan glycoprotein Rosetta Stone. New Phytol. 2, 491-500.

Lamport, D. T. A. & Várnai, P. (2013). Periplasmic arabinogalactan glycoproteins act as a calcium capacitor that regulates plant growth and development. New Phytol. 197, 58–64.

Lamport, D. T. A., Várnai, P. & Seal, C. E. (2014). Back to the future with the AGP-Ca2+ flux capacitor. Ann. Bot. 114, 1069–1085.

Lara-Espinoza, C., Carvajal-Millan, E., Balandran-Quintana, R., Lopez-Franco, Y. & Rascon-Chu, A. (2018). Pectin and pectin-based composite materials: beyond food texture. Molecules. 23(4), 942.

Le, B. H., Cheng, C., Bui, A. Q., Wagmaister, J. A., Henry, K. F., et al. (2010). Global analysis of gene activity during Arabidopsis seed development and identification of seed-specific transcription factors. Proceedings of the National Academy of Sciences. 107, 8063–8070.

Lee, C., Teng, Q., Zhong, R. & Ye, Z. H. (2012). Arabidopsis GUX proteins are glucuronyltransferases responsible for the addition of glucuronic acid side chains onto xylan. Plant Cell Physiol. 53, 1204–1216.

Lee, K. J. D., Sakata, Y., Mau, S. L., Pettolino, F., Bacic, A., et al. (2005). Arabinogalactan proteins are required for apical cell extension in the moss Physcomitrella patens. Plant Cell. 17, 3051–65.

Leonard, R., Petersen, B. O., Himly, M., Kaar, W., Wopfner, N., et al. (2005). Two novel types of O-glycans on the mugwort pollen allergen Art v 1 and their role in antibody binding. J. Biol. Chem. 280, 7932–40.

Leroux, J., Langendorff, V., Schick, G., Vaishnav, V. & Mazoyer, J. (2003). Emulsion stabilizing properties of pectin. Food Hydrocoll. 17, 455–462.

Letunic, I. & Bork, P. (2016). Interactive Tree Of Life (iTOL): An online tool for phylogenetic tree display and annotation. Bioinformatics. 23, 127–128.

Levitin, B., Richter, D., Markovich, I. & Zik, M. (2008). Arabinogalactan proteins 6 and 11 are required for stamen and pollen function in Arabidopsis. Plant J. 56, 351–363.

Li, J., Yu, M., Geng, L. L. & Zhao, J. (2010). The fasciclin-like arabinogalactan protein gene, FLA3, is involved in microspore development of Arabidopsis. Plant J. 64, 482–97.

Li, W. L., Liu, Y. & Douglas, C. J. (2017). Role of glycosyltransferases in pollen wall primexine formation and exine patterning. Plant Physiol. 173, 167–182.

Liang, Y., Basu, D., Pattathil, S., Xu, W. -L., Venetos, A., Martin, S. L., et al. (2013). Biochemical and physiological characterization of *fut4* and *fut6* mutants defective in arabinogalactan-protein fucosylation in *Arabidopsis*. J. Exp. Bot. 64, 5537–5551.

Lin, S., Dong, H., Zhang, F., Qiu, L., Wang, F. et al. (2014). *BcMF8*, a putative arabinogalactan protein-encoding gene, contributes to pollen wall development, aperture formation and pollen tube growth in *Brassicacampestris*. Ann. Bot. 113, 777–788.

Lin, S., Huang, L. & Miao, Y. (2019). Constitutive overexpression of the classical arabinogalactan protein gene *BcMF18* in *Arabidopsis* causes defects in pollen intine morphogenesis. Plant Growth Regul. 88, 159–171.

Liu, L., Shang-Guan, K., Zhang, B., Liu, X., Yan, M. et al. (2013). Brittle Culm1, a COBRA-like protein, functions in cellulose assembly through binding cellulose microfibrils. PLoS Genet. 9, e1003704.

Liu, X., Wu, X., Adhikari, P.B., Zhu, S., Kinoshita, Y. et al. (2019). Establishment of a novel method for the identification of fertilization defective mutants in *Arabidopsis thaliana*. Biochem Biophys Res Commun. 204, 928-932.

Liu, Z., Schneider, R., Kesten, C., Zhang, Y., Somssich, M. et al. (2016). Cellulose-microtubule uncoupling proteins prevent lateral displacement of microtubules during cellulose synthesis in Arabidopsis. Dev. Cell. 38, 305–315.

Lopez-Hernandez, F., Tryfona, T., Rizza, A., Yu, X. L., Harris, M. O. B. et al. (2020). Calcium binding by arabinogalactan polysaccharides is important for normal plant development. Plant Cell. 32, 3346-3369.

Lund, C. H., Bromley, J. R., Stenbæk, A., Rasmussen, R. E., Scheller, H. V. et al. (2015). A reversible Renilla luciferase protein complementation assay for rapid identification of protein-protein interactions reveals the existence of an interaction network involved in xyloglucan biosynthesis in the plant Golgi apparatus. J. Exp. Bot. 66, 85–97.

Lynch, M. & Conery, J. S. (2000). The evolutionary fate and consequences of duplicate genes. Science. 290(5494), 1151–1155.

Ma, X., Wu, Y. & Zhang, G. (2021). Formation pattern and regulatory mechanisms of pollen wall in Arabidopsis. J. Plant Physiol. 260, 153388.

MacAlister, C. A., Ortiz-Ramírez, C., Becker, J. D., Feijó, J. A. & Lippman, Z. B. (2016). Hydroxyproline O-arabinosyltransferase mutants oppositely alter tip growth in *Arabidopsis thaliana* and *Physcomitrella patens*. Plant J. 85: 193–208.

Marquez, J., Seoane-Camba, J.A. & Suarez-Cervera, M. (1997). The role of the intine and cytoplasm in the activation and germination processes of Poaceae pollen grains. Grana. 36, 328–342.

Maruyama, D., Hamamura, Y., Takeuchi, H., Susaki, D., Nishimaki, M. et al. (2013). Independent control by each female gamete prevents the attraction of multiple pollen tubes. Dev.Cell. 25, 317–323.

Maruyama, D., Völz, R., Takeuchi, H., Mori, T., Igawa, T. et al. (2015). Rapid elimination of the persistent synergid through a cell fusion mechanism. Cell. 4, 907-18.

McCarthy, T. W., Der, J. P., Honaas, L. A., dePamphilis, C. W. & Anderson, C. T. (2014). Phylogenetic analysis of pectin-related gene families in Physcomitrella patens and nine other plant species yields evolutionary insights into cell walls. B. M. C. Plant Biol. 14, 79.

McCartney, L. & Knox, J. P. (2002). Regulation of pectic polysaccharide domains in relation to cell development and cell properties in the pea testa. J. Exp. Bot. 53, 707–713.

McCormick, S. (2004) Control of male gametophyte development. Plant Cell. 16, 142–153.

McKinley, B., Rooney, W., Wilkerson, C. & Mullet, J. (2016). Dynamics of biomass partitioning, stem gene expression, cell wall biosynthesis, and sucrose accumulation during development of Sorghum bicolor. Plant. J. 88(4), 662-680.

Meents, M. J., Motani, S., Mansfield, S. D. & Samuels, A. L. (2019). Organization of xylan production in the Golgi during secondary cell wall biosynthesis. Plant Physiol. 181, 527–546.

Mendu, V., Griffiths, J. S., Persson, S., Stork, J., Downie, A. B., et al. (2011). Subfunctionalization of cellulose synthases in seed coat epidermal cells mediates secondary radial wall synthesis and mucilage attachment. Plant Physiol. 157, 441–453.

Minic, Z. (2008). Physiological role of plant glycoside hydrolases. Planta. 227, 723.

Mizukami, A. G., Inatsugi, R., Jiao, J., Kotake, T., Kuwata, K. et al. (2016). The AMOR arabinogalactan sugar chain induces pollen-tube competency to respond to ovular guidance. Curr. Biol. C. B. 26, 1091–1097.

Mizuta, Y. & Higashiyama, T. (2018). Chemical signaling for pollen tube guidance at a glance. J. Cell Sci. 131, 208447.

Mohnen, D. (2008). Pectin structure and biosynthesis. Curr. Opin. Plant Biol. 11, 266–277.

Møller, S.R., Yi, X., Velásquez, S.M., Gille, S., Hansen, P. L. M. et al. (2017). Identification and evolution of a plant cell wall specific glycoprotein glycosyl transferase. ExAD. Scientific Reports. 7, 45341.

Mori, T., Kuroiwa, H., Higashiyama, T. & Kuroiwa, T. (2006). Generative Cell Specific 1 is essential for angiosperm fertilization. Nature Cell Biology. 1, 64-71.

Mortimer, J. C., Miles, G. P., Brown, D. M., Zhang, Z., Segura, M. P. et al. (2010). Absence of branches from xylan in Arabidopsis gux mutants reveals potential for simplification of lignocellulosic biomass. Proc. Natl. Acad. Sci. U. S. A. 107, 17409–17414.

Mutwil, M., Debolt, S. & Persson, S. (2008). Cellulose synthesis: a complex complex. Curr. Opin. Plant Biol. 11, 252–57.

Nakamura, A., Furuta, H., Maeda, H., Nagamatsu, Y. & Yoshimoto, A. (2001). Analysis of structural components and molecular construction of soybean soluble polysaccharides by stepwise enzymatic degradation. Biosci. Biotechnol. Biochem. 65, 2249–2258.

Ndeh, D., Rogowski, R., Cartmell, A., Luis, A. S., Baslé, A. et al. (2017). Complex pectin metabolism by gut bacteria reveals novel catalytic functions. Nature. 544, 65–70.

Nielsen, R. & Yang, Z. (1998). Likelihood models for detecting positively selected amino acid sites and applications to the HIV-1 envelope gene. Genetics. 148, 929–936.

Øbro, J., Harholt, J., Scheller, H. V. & Orfila, C. (2004). Rhamnogalacturonan I in *Solanum tuberosum* tubers contains complex arabinogalactan structures. Phytochemistry. 65, 1429–1438.

O'Neill, M. A., Ishii, T., Albersheim, P. & Darvill, A. G. (2004). Rhamnogalacturonan II: structure and function of a borate cross-linked cell wall pectic polysaccharide. Annu. Rev. Plant Biol. 55, 109–139.

Ogawa-Ohnishi, M. & Matsubayashi, Y. (2015). Identification of three potent hydroxyproline O-galactosyltransferases in Arabidopsis. Plant J. 81, 736–746.

Ogawa-Ohnishi, M., Matsushita, W. & Matsubayashi, Y. (2013). Identification of three hydroxyproline O-arabinosyltransferases in *Arabidopsis thaliana*. Nat. Chem. Biol. 9, 726–730.

Pabst, M., Fischl, R. M., Brecker, L., Morelle, W., Fauland, A. et al. (2013). Rhamnogalacturonan II structure shows variation in the side chains monosaccharide composition and methylation status within and across different plant species. Plant J. 76, 61–72.

Paredez, A. R., Somerville, C. R. & Ehrhardt, D. W. (2006). Visualization of cellulose synthase demonstrates functional association with microtubules. Science. 312, 1491–1495.

Parsons, H. T., Stevens, T. J., McFarlane, H. E., Vidal-Melgosa, S., Griss, J. et al. (2019). Separating Golgi proteins from cis to trans reveals underlying properties of cisternal localization. Plant Cell. 31, 2010–2034.

Patel, S. & Goyal, A. (2015). Applications of Natural Polymer Gum Arabic: A Review, International Journal of Food Properties. 18:5, 986-998.

Pattathil, S., Avci, U., Baldwin, D., Swennes, A. G., McGill, J. A. et al. (2010). A comprehensive toolkit of plant cell wall glycan-directed monoclonal antibodies. Plant Physiol. 153, 514–525.

Pereira, A. M., Lopes, A. L. & Coimbra, S. (2016). *JAGGER*, an AGP essential for persistent synergid degeneration and polytubey block in Arabidopsis. Plant Signaling & Behavior. 10, 1209616.

Pereira, A. M., Masiero, S., Nobre, M. S., Costa, M. L., Solís, M. T. et al. (2014). Differential expression patterns of arabinogalactan proteins in *Arabidopsis thaliana* reproductive tissues. J. Exp. Bot. 65, 5459–5471.

Pereira, A. M., Nobre, M. S., Pinto, S. C., Lopes, A. L., Costa, M. L. et al. (2016). "Love Is Strong, and You're so Sweet": JAGGER is essential for persistent synergid degeneration and polytubey block in *Arabidopsis thaliana*. Mol. Plant. 9, 601–614.

Persson, S., Caffall, K. H., Freshour, G., Hilley, M. T., Bauer, S. et al. (2007). The Arabidopsis irregular xylem8 mutant is deficient in glucuronoxylan and homogalacturonan, which are essential for secondary cell wall integrity. Plant Cell. 19, 237–255.

Persson, S., Paredez, A., Carroll, A., Palsdottir, H., Doblin, M. et al. (2007). Genetic evidence for three unique components in primary cell-wall cellulose synthase complexes in Arabidopsis. Proc. Natl. Acad. Sci. U. S. A. 104, 15566–15571.

Potikha, T. & Delmer, D. P. (1995). A mutant of *Arabidopsis thaliana* displaying altered patterns of cellulose deposition. Plant J. 7, 453–460.

Purushotham, P., Ho, R. & Zimmer, J. (2020). Architecture of a catalytically active homotrimeric plant cellulose synthase complex. Science. 369, 1089–1094.

Qi, X., Behrens, B. X., West, P. R. & Mort, A. J. (1995). Solubilization and partial characterization of extensin fragments from cell walls of cotton suspension cultures: evidence for a covalent cross-link between extensin and pectin. Plant Physiol. 108, 1691–1701.

Qin, Y. & Zhao, J. (2006). Localization of arabinogalactan proteins in egg cells, zygotes, and two-celled proembryos and effects of b-D-glucosyl Yariv reagent on egg cell fertilization and zygote division in Nicotiana tabacum L. J. Exp. Bot. 57, 2061–2074.

Quilichini, T. D., Grienenberger, E. & Douglas, C. J. (2015). The biosynthesis, composition and assembly of the outer pollen wall: a tough case to crack. Phytochemistry. 113, 170–182.

Ragauskas, A.J., Beckham, G. T., Biddy, M. J., Chandra, R., Chen, F. et al. (2014). Lignin valorization: Improving lignin processing in the biorefinery. Science. 344, 1246843.

Ralet, M. C., Andre-Leroux, G., Quemener, B. & Thibault, J. F. (2005). Sugar beet (Beta vulgaris) pectins are covalently cross-linked through diferulic bridges in the cell wall. Phytochemistry. 66, 2800–2814.

Ralet, M. C., Crepeau, M. J., Vigouroux, J., Tran, J., Berger, A. et al. (2016). Xylans provide the structural driving force for mucilage adhesion to the Arabidopsis seed coat. Plant Physiol. 171, 165-178.

Rautengarten, C., Ebert, B., Moreno, I., Temple, H., Herter, T. et al. (2014). The Golgi localized bifunctional UDP-rhamnose/UDP-galactose transporter family of Arabidopsis. Proc. Natl. Acad. Sci. U. S. A. 111, 11563–11568.

Rautengarten, C., Usadel, B., Neumetzler, L., Hartmann, J., Büssis, D. et al. (2008). A subtilisin-like serine protease essential for mucilage release from Arabidopsis seed coats. Plant Journal. 54, 466–480.

Reddy, M. S. S., Chen, F., Shadle, G., Jackson, L., Aljoe, H. et al. (2005). Targeted down-regulation of cytochrome P450 enzymes for forage quality improvement in alfalfa (Medicago sativa L.). Proc. Natl. Acad. Sci. U. S. A. 102, 16573–16578.

Regan, S. M. & Moffatt, B. A. (1990). Cytochemical analysis of pollen development in wild type *Arabidopsis* and a male-sterile mutant. Plant Cell. 2, 877–889.

Ringli, C. (2010). The hydroxyproline-rich glycoprotein domain of the Arabidopsis LRX1 requires Tyr for function but not for insolubilization in the cell wall. Plant J. 63, 662–669.

Rui, Y. & Dinneny, J. R. (2020). A wall with integrity: surveillance and maintenance of the plant cell wall under stress. New Phytol. 225(4), 1428-1439.

Saez-Aguayo, S., Ralet, M. C., Berger, A., Botran, L., Ropartz, D. et al. (2013). PECTIN METHYLESTERASE INHIBITOR6 promotes Arabidopsis mucilage release by limiting methylesterification of homogalacturonan in seed coat epidermal cells. Plant Cell. 25, :308–323.

Saez-Aguayo, S., Rondeau-Mouro, C., Macquet, A., Kronholm, I., Ralet, M. C. et al. (2014). Local evolution of seed flotation in *Arabidopsis*. PLoS Genetics. 10, e1004221.

Saito, F., Suyama, A., Oka, T., Yoko-o, T., Matsuoka, K. et al. (2014). Identification of novel peptidyl serine α-galactosyltransferase gene family in plants. J. Biol. Chem. 289, 20405–20420.

Sampathkumar, A., Gutierrez, R., McFarlane, H. E., Bringmann, M., Lindeboom, J. et al. (2013). Patterning and lifetime of plasma membrane-localized cellulose synthase is dependent on actin organization in Arabidopsis interphase cells. Plant Physiol. 162, 675–688.

Sanchez-Rodriguez, C., Bauer, S., Hematy, K., Saxe, F., Ibanez, A. B. et al. (2012). Chitinase-like1/POM-POM1 and its homolog CTL2 are glucan-interacting proteins important for cellulose biosynthesis in Arabidopsis. Plant Cell. 24, 589-607.

Sanders, P.M., Biu, A.Q., Weterings, K., McIntire, K.N., Hsu, Y.C. et al. (1999). Anther development defects in *Arabidopsis thaliana* male-sterile mutants. Sex Plant Reprod. 11, 297-322.

Scheller, H. V. & Ulvskov, P. (2010). Hemicelluloses. Annu. Rev. Plant Biol. 61, 263–289.

Schnurr, J. A., Storey, K. K., Jung, H. J. G., Somers, D. A. & Gronwald, J. W. (2006). UDP-sugar pyrophosphorylase is essential for pollen development in Arabidopsis. Planta. 224, 520–532.

Schultink, A., Liu, L., Zhu, L. & Pauly, M. (2014). Structural diversity and function of xyloglucan sidechain substituents. Plants. 3, 526–542.

Schultz, C. J., Johnson, K. L., Currie, G. & Bacic, A. (2000). The classical arabinogalactan protein gene family of Arabidopsis. Plant Cell. 12, 1751–68.

Schultz, C. J., Rumsewicz, M. P., Johnson, K. L., Jones, B. J., Gaspar, Y. M. et al. (2002). Using genomic resources to guide research directions. The arabinogalactan protein gene family as a test case. Plant Physiol. 129, 1448–63.

Scimeca, M., Bischetti, S., Lamsira, H. K., Bonfiglio, R. & Bonanno, E. (2018). Energy Dispersive X-ray (EDX) microanalysis: A powerful tool in biomedical research and diagnosis. Eur. J. Histochem. 62, 2841.

Scott, R. J., Spielman, M. & Dickinson, H. D. (2004). Stamen structure and function. Plant Cell. 16, 46–60.

Seifert, G. J. (2020). On the potential function of type II arabinogalactan O-glycosylation in regulating the fate of plant secretory proteins. Front Plant Sci. 11, 563735.

Seifert, G. J. & Roberts, K. (2007). The biology of arabinogalactan proteins. Annu Rev Plant Biol. 58, 137-61.

Selvendran, R. R. & Ryden, P. (1990). Isolation and analysis of plant cell walls. In Methods in Plant Biochemistry, Vol. 2: Carbohydrates, P.M. Dey and J.B. Harbourne, eds (San Diego, CA: Academic Press) pg. 549–575.

Shi, H., Kim, Y., Guo, Y., Stevenson, B. & Zhu, J. K. (2003). The Arabidopsis SOS5 locus encodes a putative cell surface adhesion protein and is required for normal cell expansion. Plant Cell. 15, 19–32.

Shi, J., Cui, M., Yang, L., Kim, Y. J. & Zhang, D. (2015). Genetic and biochemical mechanisms of pollen wall development. Trends Plant Sci. 20, 741–753.

Shimoda, R., Okabe, K., Kotake, T., Matsuoka, K., Koyama, T. et al. (2014). Enzymatic fragmentation of carbohydrate moieties of radish arabinogalactan-protein and elucidation of the structures. Biosci. Biotechnol. Biochem. 78, 818–831.

Showalter, A. M. (1993). Structure and function of plant cell wall proteins. Plant Cell. 5, 9–23.

Showalter, A. M. (2001). Arabinogalactan-proteins: structure, expression and function. Cell Mol. Life Sci. 58, 1399–1417.

Showalter, A. M. & Basu, D. (2016). Extensin and arabinogalactan-protein biosynthesis: glycosyltransferases, research challenges, and biosensors. Front Plant Sci. 7, 814.

Showalter, A. M., Keppler, B., Lichtenberg, J., Gu, D. & Welch, L. R. (2010). A bioinformatics approach to the identification, classification, and analysis of hydroxyproline-rich glycoproteins. Plant Physiol. 153, 485–513.

Shpak, E., Barbar, E., Leykam, J. F. & Kieliszewski, M. J. (2001). Contiguous hydroxyproline residues direct hydroxyproline arabinosylation in Nicotiana tabacum. J. Biol. Chem. 276, 11272–11278.

Shpak, E., Leykam, J. F. & Kieliszewski, M. J. (1999). Synthetic genes for glycoprotein design and the elucidation of hydroxyproline-O-glycosylation codes. Proc. Natl. Acad. Sci. U.S.A. 96(26), 14736–14741.

Simmons, B. A., Logue, D. & Ralph, J. (2010). Advances in modifying lignin for enhanced biofuel production. Curr Opin Plant Biol. 13, 313-320.

Sindhu, A., Langewisch, T., Olek, A., Multani, D. S., McCann, M. C. et al. (2007). Maize brittle stalk2 encodes a COBRA-like protein expressed in early organ development but required for tissue flexibility at maturity. Plant Physiol. 145, 1444–1459.

Smith, P. J., Wang, H. T., York, W. S., Pena, M. J. & Urbanowicz, B. R. (2017). Designer biomass for next-generation biorefineries: Leveraging recent insights into xylan structure and biosynthesis. Biotechnol. Biofuels. 10, 286.

Smyth, D. R., Bowman, J. L. &and Meyerowitz, E. M. (1990). Early flower development in *Arabidopsis*. Plant Cell. 2, 755–767.

Sommer-Knudsen, J., Lush, W. M., Bacic, A. & Clarke, A. E. (1998). Re-evaluation of the role of a transmitting tract-specific glycoprotein on pollen tube growth. Plant J. 13, 529–35.

Sørensen, I., Pettolino, F. A., Bacic, A., Ralph, J., Lu, F. et al. (2011). The charophycean green algae provide insights into the early origins of plant cell walls. Plant J. 68, 201–211.

Stork, J., Harris, D., Griffiths, J., Williams, B., Beisson, F. et al. (2010). CELLULOSE SYNTHASE 9 serves a nonredundant role in secondary cell wall synthesis in Arabidopsis epidermal testa cells. Plant Physiol. 153, 580–589.

Strasser, R. (2016). Plant protein glycosylation. Glycobiology, 26(9), 926–939.

Strasser, R., Bondili, J. S., Vavra, U., Schoberer, J., Svoboda, B. et al. (2007). A unique beta1,3-galactosyltransferase is indispensable for the biosynthesis of N-glycans containing Lewis a structures in *Arabidopsis thaliana*. Plant Cell. 19, 2278–2292.

Sullivan, S., Ralet, M. C., Berger, A., Diatloff, E., Bischoff, V. et al. (2011). CESA5 is required for the synthesis of cellulose with a role in structuring the adherent mucilage of Arabidopsis seeds. Plant Physiol. 156, 1725–1739.

Suzuki, T., Masaoka, K., Nishi, M., Nakamura, K. & Ishiguro, S. (2008). Identification of kaonashi mutants showing abnormal pollen exine structure in Arabidopsis thaliana. Plant Cell Physiol. 49, 1465–1477.

Suzuki, T., Narciso, J. O., Zeng, W., van de Meene, A., Yasutomi, M. et al. (2017). KNS4/UPEX1: A Type II arabinogalactan β-(1,3)-galactosyltransferase required for pollen exine development. Plant Physiol. 173, 183–205.

Takenaka, Y., Kato, K., Ogawa-Ohnishi, M., Tsuruhama, K., Kajiura, H. et al. (2018). Pectin RG-I rhamnosyltransferases represent a novel plant-specific glycosyltransferase family. Nat. Plants. 4, 669–676.

Takeuchi, H. & Higashiyama, T. (2012). A species-specific cluster of defensin-like genes encodes diffusible pollen tube attractants in Arabidopsis. PLoS Biol. 10, 1001449.

Talbot, M. J. & White, R. G. (2013). Methanol fixation of plant tissue for Scanning Electron Microscopy improves preservation of tissue morphology and dimensions. Plant Methods. 9, 36.

Tan, L., Eberhard, S., Pattathil, S., Warder, C., Glushka, J. et al. (2013). An Arabidopsis cell wall proteoglycan consists of pectin and arabinoxylan covalently linked to an arabinogalactan protein. Plant Cell. 25, 270–287.

Tan, L., Leykam, J. F. & Kieliszewski, M. J. (2003). Glycosylation motifs that direct arabinogalactan addition to arabinogalactan-proteins. Plant Physiol. 132, 1362–69.

Tan, L., Showalter, A. M., Egelund, J., Hernandez-Sanchez, A., Doblin, M. S. et al. (2012). Arabinogalactan-proteins and the research challenges for these enigmatic plant cell surface proteoglycans. Front. Plant Sci. 3, 140.

Tan, L., Qiu, F., Lamport, D. T. & Kieliszewski, M. J. (2004). Structure of a

hydroxyproline (Hyp)-arabinogalactan polysaccharide from repetitive Ala-Hyp

expressed in transgenic Nicotiana tabacum. J. Biol. Chem. 279, 13156–65.

Tan, L., Varnai, P., Lamport, D. T. A., Yuan, C. H., Xu, J. F. et al. (2010). Plant O–

hydroxyproline arabinogalactans are composed of repeating trigalactosyl subunits

with short, bifurcated side chains. J. Biol. Chem. 285, 24575–24583.

Tautz, D. & Domazet-Loso, T. (2011). The evolutionary origin of orphan genes. Nat.

Rev. Genet. 12(10), 692–702.

Taylor, N. G., Howells, R. M., Huttly, A. K., Vickers, K. & Turner, S. R. (2003).

Interactions among three distinct CesA proteins essential for cellulose synthesis.

Proc. Natl. Acad. Sci. U. S. A. 100, 1450–1455.

Temple, H., Mortimer, J. C., Tryfona, T., Yu, X., Lopez-Hernandez, F. et al. (2019). Two

members of the DUF579 family are responsible for arabinogalactan methylation

in Arabidopsis. Plant Direct. 3, 1–4.

Thimm, J. C., Burritt, D. J., Sims, I. M., Newman, R. H., Ducker, W. A. et al. (2002).

Celery (Apium graveolens) parenchyma cell walls: cell walls with minimal

xyloglucan. Physiol. Plant. 116, 164–171.

Thimmaiah, C., Shetty, P., Shetty, S. B., Natarajan S. & Thomas, N. (2019). A

Comparative analysis of the remineralization potential of CPP–ACP with

Fluoride, Tri-Calcium Phosphate and Nano Hydroxyapatite using SEM/EDX–An

in vitro study. J. Clin. Exp. Dent. 11, 1120-1126.

Thomas, L. H., Forsyth, V. T., Martel, A., Grillo, I., Altaner, C. M. et al. (2015). Diffraction evidence for the structure of cellulose microfibrils in bamboo, a model for grass and cereal celluloses. B. M. C. Plant Biol. 15, 153.

Thor, K. (2019). Calcium—Nutrient and Messenger. Front. Plant Sci. 10, 440.

Tiainen, P., Myllyharju, J. & Koivunen, P. (2005). Characterization of a second *Arabidopsis thaliana* prolyl 4-hydroxylase with distinct substrate specificity. J. Biol. Chem. 280, 1142–8.

Tryfona, T., Liang, H. C., Kotake, T., Kaneko, S., Marsh, J. et al. (2010). Carbohydrate structural analysis of wheat flour arabinogalactan protein. Carbohydr. Res. 345, 2648–2656.

Tryfona, T., Liang, H. C., Kotake, T., Tsumuraya, Y., Stephens, E. et al. (2012). Structural characterization of Arabidopsis leaf arabinogalactan polysaccharides. Plant Physiol. 160, 653–666.

Tryfona, T., Theys, T. E. Wagner, T., Stott, K., Keegstra, K. et al. (2014). Characterisation of FUT4 and FUT6 α-(1−→2)-fucosyltransferases reveals that absence of root arabinogalactan fucosylation increases Arabidopsis root growth salt sensitivity. PLoS ONE. 9, e93291.

Tuomivaara, S. T., Yaoi, K., O'Neill, M. A. & York, W. S. (2015). Generation and structural validation of a library of diverse xyloglucan-derived oligosaccharides, including an update on xyloglucan nomenclature. Carbohydr. Res. 402, 56–66.

Twell, D. (2010) Male gametophyte development. In: Pua EC, Davey MR (eds) Plant developmental biology-biotechnological perspectives. Springer, Berlin. 1, 225–244.

Uemura, T., Nakano, R. T., Takagi, J., Wang, Y., Kramer, K. et al. (2019). A Golgi-released subpopulation of the trans-Golgi network mediates protein secretion in Arabidopsis. Plant Physiol. 179, 519–532.

Updegraff, D. M. (1969). Semi-micro determination of cellulose in biological materials. Anal Biochem. 32, 420–424.

Urbanowicz, B. R., Pena, M. J., Moniz, H. A., Moremen, K. W. & York, W. S. (2014). Two Arabidopsis proteins synthesize acetylated xylan in vitro. Plant J. 80, 197–206.

Vain, T., Crowell, E. F., Timpano, H., Biot, E., Desprez, T. et al. (2014). The cellulase KORRIGAN is part of the cellulose synthase complex. Plant Physiol. 165, 1521–1532.

Vamathevan, J. J., Hasan, S., Emes, R. D., Amrine-Madsen, H., Rajagopalan, D. et al. (2008). The role of positive selection in determining the molecular cause of species differences in disease. BMC. Evol. Biol. 8, 273.

van Hengel, A. J. & Roberts, K. (2002). Fucosylated arabinogalactan-proteins are required for full root cell elongation in Arabidopsis. Plant J. 32, 105–113.

van Hengel, A. J. & Roberts, K. (2003). AtAGP30, an arabinogalactan-protein in the cell walls of the primary root, plays a role in root regeneration and seed germination. Plant J. 36, 256–70.

Vaahtera, L., Schulz, J. & Hamann, T. (2019). Cell wall integrity maintenance during plant development and interaction with the environment. Nat. Plants. 5, 924–932.

Vanneste, S. & Friml, J. (2013). Calcium: the missing link in auxin action. Plants. 2, 650–675.

Varidenbosch, K. A., Bradley, D. J., Knox, J. P., Perotto, S., Butcher, G. W. et al. (1989). Common components of the infection thread matrix and the intercellular space identified by immunocytochemical analysis of pea nodules and uninfected roots. E. M. B. O. J. 8, 335-341.

Vega-Sanchez, M. E., Verhertbruggen, Y., Christensen, U., Chen, X., Sharma, V. et al. (2012). Loss of cellulose synthase-like F6 function affects mixed-linkage glucan deposition, cell wall mechanical properties, and defense responses in vegetative tissues of rice. Plant Physiol. 159, 56–69.

Velasquez, S. M., Marzol, E., Borassi, C., Pol-Fachin, L., Ricardi, M. M. et al. (2015). Low sugar is not always good: Impact of specific O-glycan defects on tip growth in Arabidopsis. Plant Physiol. 168, 808–813.

Velasquez, S. M., Ricardi, M. M., Gloazzo Dorosz, J., Fernandez, P. V., Nadra, A. D. et al. (2011). O-glycosylated cell wall extensins are essential in root hair growth Science. 33, 1401–1403.

Velasquez, M., Salter, J. S., Dorosz, J. G., Petersen, B. L. & Estevez, J. M. (2012). Recent advances on the posttranslational modifications of extensins and their roles in plant cell walls. Front. Plant Sci. 3, 93.

Verger, S., Chabout, S., Gineau, E. & Mouille, G. (2016). Cell adhesion in plants is under the control of putative O-fucosyltransferases. Development. 143, 2536–2540.

Viotti, C., Bubeck, J., Stierhof, Y. D., Krebs, M., Langhans, M. et al. (2010). Endocytic and secretory traffic in Arabidopsis merge in the trans-Golgi network/early endosome, an independent and highly dynamic organelle. Plant Cell. 22, 1344–1357.

Vlad, F., Spano, T., Vlad, D., Bou Daher, F., Ouelhadj, A. et al. (2007). Arabidopsis prolyl 4-hydroxylases are differentially expressed in response to hypoxia, anoxia and mechanical wounding. Physiol. Plant. 130, 471–483.

Voiniciuc, C., Dean, G. H., Griffiths, J. S., Kirchsteiger, K., Hwang, Y. T. et al. (2013). Flying Saucer1 is a transmembrane RING E3 ubiquitin ligase that regulates the degree of pectin methylesterification in Arabidopsis seed mucilage. The Plant Cell. 25, 944-959.

Voiniciuc, C., Engle, K. A., Günl, M., Dieluweit, S., Schmidt, M. H. W. et al. (2018). Identification of key enzymes for pectin synthesis in seed mucilage. Plant Physiol. 178, 1045–1064.

Voiniciuc, C., Gunl, M., Schmidt, M. H. W. & Usadel, B. (2015). Highly branched xylan made by IRX14 and MUCI21 links mucilage to Arabidopsis seeds. Plant Physiol. 169, 2481–2495.

Voiniciuc, C., Schmidt, M. H., Berger, A., Yang, B., Ebert, B. et al. (2015). MUCILAGE RELATED10 produces galactoglucomannan that maintains pectin and cellulose architecture in Arabidopsis seed mucilage. Plant Physiol. 169, 403–420.

Voiniciuc, C., Yang, B., Schmidt, M. H. W., Günl, M., & Usadel, B. (2015). Starting to gel: how arabidopsis seed coat epidermal cells produce specialized secondary cell walls. Int. J. Mol. Sci. 16, 3452–3473.

Völz, R., Heydlauff, J., Ripper, D., von Lyncker, L. & Groß-Hardt, R. (2013) Ethylene signaling is required for synergid degeneration and the establishment of a pollen tube block. Dev. Cell. 25, 310–316.

Voorrips, R. E. (2002). MapChart: Software for the graphical presentation of linkage maps and QTLs. J. Hered. 93, 77–78.

Waldron, K. W., Parker, M. L. & Smith, A. C. (2003). Plant cell walls and food quality. Comprehensive Reviews in Food Science and Food Safety. 2(4), 128–146.

Waterhouse, A. M., Proctor, J. B., Martin, D. M. A., Clamp, M., & Barton, G. J. (2009). Jalview Version 2—a multiple sequence alignment editor and analysis workbench. Bioinformatics 25, 1189–1191.

Western, T. L., Skinner, D. J., & Haughn, G. W. (2000). Differentiation of mucilage secretory cells of the Arabidopsis seed coat. Plant Physiol. 122, 345-356.

Western, T. L., Young, D. S., Dean, G. H., Tan, W. L., Samuels, A. L. et al. (2004). MUCILAGE-MODIFIED4 encodes a putative pectin biosynthetic enzyme developmentally regulated by APETALA2, TRANSPARENT TESTA GLABRA1, and GLABRA2 in the Arabidopsis seed coat. Plant Physiol. 134, 296–306.

Whelan, S. G. N. (1999). Distributions of statistics used for the comparison of models of sequence evolution in phylogenetics. Mol. Biol. Evol. 19, 1292.

Wiggins, C. A. R. & Munro, S. (1998). Activity of the yeast MNN1 α-1,3-mannosyltransferase requires a motif conserved in many other families of glycosyltransferases. P. N. A. S. 95(14), 7845–7950.

Willats, W. G. T., McCartney, L. & Knox, J. P. (2001). In-situ analysis of pectic polysaccharides in seed mucilage and at the root surface of Arabidopsis thaliana. Planta. 213, 37–44.

Williamson, R. E., Burn, J. E., Birch, R., Baskin, T. I., Arioli, T. et al. (2001). Morphology of rsw1, a cellulose-deficient mutant of *Arabidopsis thaliana*. Protoplasma. 215, 116–127.

Wilson, S. M., Ho, Y. Y., Lampugnani, E. R., Van de Meene, A. M., Bain, M. P. et al. (2015). Determining the subcellular location of synthesis and assembly of the cell wall polysaccharide (1,3; 1,4)-β-D-glucan in grasses. Plant Cell. 27, 754–771.

Winter, D., Vinegar, B., Nahal, H., Ammar, R., Wilson, G. V. et al. (2007). An "Electronic Fluorescent Pictograph" browser for exploring and analyzing large-scale biological data sets. PLoS One. 2, e718.

Wolf, S. (2017). Plant cell wall signalling and receptor-like kinases. Biochemical Journal. 474, 471–492.

Wong, W. S., Yang, Z., Goldman, N. & Nielsen, R. (2004). Accuracy and power of statistical methods for detecting adaptive evolution in protein coding sequences and for identifying positively selected sites. Genetics. 168, 1041–1051.

Wu, A. M., Hornblad, E., Voxeur, A., Gerber, L., Rihouey, C. et al. (2010). Analysis of the Arabidopsis IRX9/IRX9-L and IRX14/IRX14-L pairs of glycosyltransferase genes reveals critical contributions to biosynthesis of the hemicellulose glucuronoxylan. Plant Physiol. 153, 542–554.

Wu, B., Zhang, B., Dai, Y., Zhang, L., Shang-Guan, K. et al. (2012). Brittle culm15 encodes a membrane-associated chitinase-like protein required for cellulose biosynthesis in rice. Plant Physiol. 159, 1440–1452.

Wu, H., Wang, H. & Cheung, A. Y. (1995). A pollen tube growth stimulatory glycoprotein is deglycosylated by pollen tubes and displays a glycosylation gradient in the flower. Cell. 82, 395–403.

Wu, Y., Williams, M., Bernard, S., Driouich, A., Showalter, A. M. et al. (2010). Functional identification of two nonredundant Arabidopsis alpha (1,2) fucosyltransferases specific to arabinogalactan proteins. J. Biol. Chem. 285, 13638–13645.

Xu, S. L., Rahman, A., Baskin, T. I. & Kieber, J. J. (2008). Two leucine-rich repeat receptor kinases mediate signaling, linking cell wall biosynthesis and ACC synthase in Arabidopsis. Plant Cell. 20(11), 3065–3079.

Xu, J., Tan, L., Lamport, D. T. A., Showalter, A. M. & Kieliszewski, M. J. (2008). The O-Hyp glycosylation code in tobacco and Arabidopsis and a proposed role of Hyp-glycans in secretion. Phytochemistry. 69, 1631–1640.

Yang, B., Voiniciuc, C., Fu, L., Dieluweit, S., Klose, H. et al. (2019). TRM4 is essential for cellulose deposition in Arabidopsis seed mucilage by maintaining cortical microtubule organization and interacting with CESA3. New Phytologist. 221, 881-895.

Yang, Z., Wong, W. S. & Nielsen, R. (2005). Bayes empirical Bayes inference of amino acid sites under positive selection. Mol. Biol. Evol. 22, 1107–1118.

Yapo, B. M. (2011). Rhamnogalacturonan-I: a structurally puzzling and functionally versatile polysaccharide from plant cell walls and mucilages. Polym. Rev. 51, 391–413.

Yapo, B. M., Lerouge, P., Thibault, J. F. & Ralet, M. C. (2007). Pectins from citrus peel cell walls contain homogalacturonans homogenous with respect to molar mass, rhamnogalacturonan I and rhamnogalacturonan II. Carbohydr. Polym. 69, 426–435.

Ye, C. Y., Li, T., Tuskan, G. A., Tschaplinski, T. J. & Yang, X. (2011). Comparative analysis of GT14/GT14-like gene family in Arabidopsis, Oryza, Populus, Sorghum and Vitis. Plant Sci. 181(6), 688–695.

Young, R. E., McFarlane, H. E., Hahn, M. G., Western, T. L., Haughn, G. W. et al. (2008). Analysis of the Golgi apparatus In Arabidopsis seed coat cells during polarized secretion of pectin-rich mucilage. Plant Cell. 20, 1623-1638.

Yuasa, K., Toyooka, K., Fukuda, H. & Matsuoka, K. (2005). Membrane-anchored prolyl hydroxylase with an export signal from the endoplasmic reticulum. Plant J. 41, 81–9410.

Zabotina, O. A. (2012). Xyloglucan and its biosynthesis. Front. Plant Sci. 3, 134.

Zeng, W., Ford, K. L., Bacic, A. & Heazlewood, J. L. (2018). N-linked glycan micro-heterogeneity in glycoproteins of Arabidopsis. Mol. Cell. Proteomics. 17, 413–421.

Zeng, W., Lampugnani, E. R., Picard, K. L., Song, L., Wu, A. M. et al. (2016). Asparagus IRX9, IRX10, and IRX14A are components of an active xylan backbone synthase complex that forms in the Golgi apparatus. Plant Physiol. 171, 93–109.

Zhang, B., Deng, L., Qian, Q., Xiong, G., Zeng, D. et al. (2009). A missense mutation in the transmembrane domain of CESA4 affects protein abundance in the plasma membrane and results in abnormal cell wall biosynthesis in rice. Plant Mol. Biol. 71, 509–524.

Zhang, B., Gao, Y., Zhang, L. & Zhou, Y. (2021). The plant cell wall: Biosynthesis, construction, and functions. J. Integr. Plant Biol. 63(1), 251-272.

Zhang, J., Nielsen, R. & Yang, Z. (2005). Evaluation of an improved branch-site likelihood method for detecting positive selection at the molecular level. Mol. Biol. Evol. 22, 2472–2479.

Zhang, Y., Held, M. A. & Showalter, A. M. (2020). Elucidating the roles of three β-glucuronosyltransferases (GLCATs) acting on arabinogalactan-proteins using a CRISPR-Cas9 multiplexing approach in Arabidopsis. B. M. C. Plant Biol. 20, 221.

Zhang, Y., Held, M. A., Kaur, D. & Showalter, A. M. (2021). CRISPR-Cas9 multiplex genome editing of the hydroxyproline-O-galactosyltransferase gene family alters arabinogalactan-protein glycosylation and function in Arabidopsis. B. M. C. Plant Biol. 21(1), 16.

Zhao, X., Qiao, L. & Wu, A. (2017). Effective extraction of Arabidopsis adherent seed mucilage by ultrasonic treatment. Scientific Reports. 7, 40672.

Zhong, R. Q., Cui, D. T. & Ye, Z. H. (2018). Members of the DUF231 family are O-acetyltransferases catalyzing 2-O- and 3-O-acetylation of mannan. Plant Cell Physiol. 59, 2339–2349.

Zhong, S. & Qu, L. J. (2019) Peptide/receptor-like kinase-mediated signaling involved in male-female interactions. Curr. Opin. Plant Biol. 51, 7–14.

Zhu, J. K. (2016). Abiotic stress signaling and responses in plants. Cell. 167, 313–324.

Note: The supplemental tables are numbered according to the chapter they appear.

Supplemental Table 2-1. Distribution of Putative *GLCAT* Genes Across 14 Plant Genomes. Eleven genes were identified as members of the *GLCAT* genes in *Arabidopsis thaliana*, *Arabidopsis lyrata*, *Brachipodium distachyon*, *Oryza sativa*, *Amborella trichopoda* and *Sorghum bicolor* (sorghum). Thirteen *GLCAT* genes were identified in *Citrus sinensis* (orange) and *Populus trichocarpa* (poplar), 22 in *Glycine max* (soybean), 15 in *Solanum lycopersicum* (tomato) and *Gossypium raimondi* (cotton), 6 in *Physcomitrella patens*, 2 in *Selaginella moellendorffii*, 9 in *Vitis vinifera* (grape) and 10 in *Zea mays* (corn)

Species name: *Arabidopsis lyrata*

gene name	pI	MW	chromosome location	genomic cordinates	peptide sizes
AL7G50100.t1	9.22	50865	7	scaffold_7:22746489..22748943	445
AL6G25870.t1	9.22	49494.8	6	scaffold_6:6111550..6113494	434
AL7G25290.t1	9.25	47779.7	7	scaffold_7:6197491..6199686	421
AL3G27750.t1	9.17	48337.6	3	scaffold_3:6408926..6413281	424
AL1G13030.t1	9.02	51295.8	1	scaffold_1:1039755..1042079	447
AL6G49580.t1	9.11	51828.5	6	scaffold_6:23213893..23216173	449
AL1G61460.t1	9.41	48054.6	1	scaffold_1:28290907..28293071	422
AL3G13490.t1	9.82	43838.1	3	scaffold_3:1178438..1180970	378
AL2G30650.t1	6.37	44997.5	2	scaffold_2:14774137..14776431	395
AL3G39340.t1	8.15	46908.3	3	scaffold_3:11464822..11467214	416
AL4G34150.t1	7.01	26324.1	4	scaffold_4:18035507..18037398	223

Species name: *Arabidopsis thaliana*

gene name	pI	MW	chromosome location	genomic cordinates	peptide sizes
at5g39990	9.32	51118.3	5	Chr5:16004281..16006740	447

gene name	pI	MW	chromosome location	genomic coordinates	peptide sizes
at2g37585	9.41	44459.4	2	Chr2:15765784..15767878	384
at5g15050	9.22	49583	5	Chr5:4871639..4873598	434
at1g53100	9.23	48166.7	1	Chr1:19786956..19788969	423
at1g03520	8.74	51503.8	1	Chr1:877862..880158	447
at3g15350	9.41	48541	3	Chr3:5166206..5169151	424
at3g24040	6.87	47195.6	3	Chr3:8680877..8683289	417
at1g71070	6.21	44918.4	1	Chr1:26807188..26809504	395
at3g03690	9.88	43823.1	3	Chr3:911238..913677	378
at4g03340	8.59	51667.3	4	Chr4:1467727..1470080	448
at4g27480	9.33	47811.8	4	Chr4:13736835..13738915	421

Species name: *Brachipodium distachyon*

gene name	pI	MW	chromosome location	genomic cordinates	peptide sizes
Bradi4g00410.1	6.76	46787.6	4	Bd4:166729..169662	424
Bradi1g36660.1	7.81	49528.6	1	Bd1:32438031..32442967	441
Bradi1g12030.2	6.07	55460	1	Bd1:8933713..8936483	511
Bradi3g14860.1	5.86	53731.9	3	Bd3:13183109..13187160	477
Bradi2g06240.1	9.05	52964.9	2	Bd2:4691454..4694243	483
Bradi1g66497.1	9.23	46406.2	1	Bd1:65498982..65503834	412
Bradi3g27240.2	9.94	48880.1	3	Bd3:27887744..27890788	442
Bradi2g35570.1	9.6	47441.4	2	Bd2:35667417..35670651	427
Bradi1g75410.1	6.94	46270.9	1	Bd1:72547246..72550530	425
Bradi2g01620.1	8.91	45354.7	2	Bd2:1107484..1112123	402
Bradi2g51360.1	5.27	44471	2	Bd2:50737114..50738970	413

Species name: *Citrus sinensis*

gene name	pI	MW	chromosome location	genomic cordinates	peptide sizes
orange1.1g037238m	9.41	46298.2	6	scaffold00015:724908..727490	410
orange1.1g014438m	8.27	48510.3	4	scaffold00005:2296412..2299072	424
orange1.1g015902m	7.77	45558.9	1	scaffold00005:2296412..2299072	424
orange1.1g016354m	6.87	44924.1	1	scaffold00005:2296412..2299072	424
orange1.1g013052m	9.03	51530.4	1	scaffold00482:93209..97008	450
orange1.1g015219m	9.36	47032.5	1	scaffold00314:103997..106900	411
orange1.1g015112m	9.08	46908.2	1	scaffold00047:232248..235811	413
orange1.1g015070m	9.08	46908.2	1	scaffold00047:232248..235811	413
orange1.1g015224m	9.73	46731.1	1	scaffold00029:860047..863700	411
orange1.1g015651m	9.64	47705.1	1	scaffold00029:860047..863700	411
orange1.1g039162m	8.63	46035.6	1	scaffold00034:775984..779461	406
orange1.1g012284m	8.11	52315.5	1	scaffold00538:50022..54220	467
orange1.1g014248m	8.69	48254.5	1	scaffold00468:88150..90863	428

Species name: *Glycine max*

gene name	pI	MW	chromosome location	genomic cordinates	peptide sizes
Glyma.13G065900.1	9.19	48799.1	13	Chr13:16584367..16588728	429
Glyma.19G018800.1	9.03	47647.6	19	Chr19:1974856..1977977	416
Glyma.18G254900.1	9.11	49694.1	18	Chr18:54114517..54118574	435
Glyma.13G167500.1	8.6	48247.2	13	Chr13:28196991..28200871	420
Glyma.17G115100.1	7.79	48658.5	17	Chr17:9099718..9103647	422
Glyma.04G148100.1	9.11	48537.4	4	Chr04:30262528..30271665	424
Glyma.06G222700.1	9.11	49187.2	6	Chr06:29116739..29120404	429

gene name	pI	MW	chromosome location	genomic cordinates	peptide sizes
Glyma.06G297300.1	8.87	49309.6	6	Chr06:48646058..48650679	432
Glyma.12G107700.1	8.74	49134.5	12	Chr12:9855268..9860062	432
Glyma.18G149600.1	9.31	47492	18	Chr18:26866084..26870647	415
Glyma.12G155800.1	9.27	48462.5	12	Chr12:24581219..24585195	423
Glyma.06G240700.1	9.22	48325.2	6	Chr06:39763159..39767344	422
Glyma.12G224700.1	8.79	48106.8	12	Chr12:38418439..38422071	420
Glyma.13G276600.2	9.04	48006.7	13	Chr13:37808033..37811453	420
Glyma.16G035500.1	9.87	45906.2	16	Chr16:3366835..3371529	399
Glyma.19G117600.1	9.86	45985.4	19	Chr19:37378187..37383287	399
Glyma.10G264600.1	9.21	45087.7	10	Chr10:48877329..48882553	396
Glyma.03G083000.1	9.29	44754.6	3	Chr03:23023402..23030458	395
Glyma.09G119700.1	6.32	46313.9	9	Chr09:28647611..28654489	405
Glyma.07G165800.1	6.46	46072.6	7	Chr07:25570901..25576900	405
Glyma.10G127100.1	8.21	48169.1	10	Chr10:33821417..33825886	430
Glyma.07G020500.1	6.61	48517.4	7	Chr07:1603103..1605882	423

Species name: *Gossypium raimondii*

gene name	pI	MW	chromosome location	genomic cordinates	peptide sizes
Gorai.006G025900.1	9.3	49298.8	6	Chr06:6540047..6544098	432
Gorai.007G078400.1	9.13	49227.5	7	Chr07:5572537..5575732	433
Gorai.004G277100.1	9.11	48550.6	4	Chr04:61028598..61032885	427
Gorai.008G152300.1	9.24	51095.7	8	Chr08:40950034..40954679	450
Gorai.002G017200.1	9.35	50848.4	2	Chr02:1140033..1143537	444
Gorai.007G375700.1	9.15	47940	7	Chr07:60717651..60720340	416
Gorai.013G168800.1	9.24	47777.4	13	Chr13:45450896..45454314	423
Gorai.007G323000.1	8.79	46955	7	Chr07:54089632..54092440	417

gene name	pI	MW	chromosome location	genomic cordinates	peptide sizes
Gorai.002G069800.1	9.31	47620.6	2	Chr02:8090529..8092516	418
Gorai.007G254100.1	9.89	45965.4	7	Chr07:41006723..41010839	400
Gorai.005G162900.1	8.15	45776.5	5	Chr05:47335894..47339911	399
Gorai.011G162900.2	9.72	48498	11	Chr11:30718965..30722814	461
Gorai.002G180500.1	7.14	45367.1	2	Chr02:47488037..47493135	396
Gorai.005G039100.1	8.83	48395.4	5	Chr05:3700631..3704420	434
Gorai.012G103300.1	8.97	38803.3	12	Chr12:23040797..23045926	340

Species name: *Oryza sativa*

gene name	pI	MW	chromosome location	genomic cordinates	peptide sizes
LOC_Os12g44240.1	7.31	47468.4	12	Chr12:27436231..27438509	426
LOC_Os06g40060.1	7.77	49252.3	6	Chr6:23843212..23847136	444
LOC_Os03g48560.1	7.18	48799.8	3	Chr3:27678462..27681754	449
LOC_Os08g04790.1	5.75	52176.2	8	Chr8:2389729..2393042	466
LOC_Os01g10440.1	9.17	52132.3	1	Chr1:5507002..5509548	480
LOC_Os03g16890.1	9.58	45371.2	3	Chr3:9382925..9387439	402
LOC_Os10g30080.1	10.17	47544.6	10	Chr10:15621355..15624800	420
LOC_Os01g03160.1	8.55	45460.8	1	Chr1:1232473..1236832	402
LOC_Os05g06050.1	8.46	45141.8	5	Chr5:3030924..3033213	406
LOC_Os04g23580.1	6.47	44072.8	4	Chr4:13493211..13498065	401
LOC_Os01g56570.1	5.39	44932.5	1	Chr1:32611406..32613200	419

Species name: *Physcomitrella patens*

gene name	pI	MW	chromosome location	genomic cordinates	peptide sizes
Pp3c14_2510V3.9	9.49	47306.8	14	Chr14:1748168..175304	417

gene name	pI	MW	chromosome location	genomic cordinates	peptide sizes
Pp3c10_2430V3.10	9.53	47753.2	10	Chr10:2028592..2034267	418
Pp3c19_19350V3.2	9.58	56861.2	19	Chr19:12862480..12866089	498
Pp3c17_12000V3.1	9.02	51358.3	17	Chr17:8593746..8597634	451
Pp3c26_670V3.1	8.34	50601.1	26	Chr26:275468..279379	448
Pp3c9_900V3.1	9.89	52266.4	9	Chr09:478824..482769	454

Species name: *Populus trichocarpa*

gene name	pI	MW	chromosome location	genomic cordinates	peptide sizes
Potri.017G075600.1	8.88	48996.1	17	Chr17:8859355..8863444	428
Potri.004G130300.1	9.21	48937.1	4	Chr04:14595212..14599309	428
Potri.013G145500.2	8.84	50488.7	13	Chr13:15118532..15123612	442
Potri.009G003300.1	7.75	50157.6	9	Chr09:773983..777805	438
Potri.019G104600.1	9.26	50792.3	19	Chr19:13327129..13332419	442
Potri.001G399500.2	9.24	47754.2	1	Chr01:42111736..42115574	419
Potri.011G119600.1	9.14	47235.3	11	Chr11:14492700..14496253	419
Potri.006G263000.1	8.61	51327.2	6	Chr06:26582875..26585580	448
Potri.013G066200.1	9.85	45660.9	13	Chr13:5087915..5092450	397
Potri.008G128000.1	8.43	45261	8	Chr08:8341015..8345946	396
Potri.010G114800.3	8.46	45136.8	10	Chr10:13302987..13308442	400
Potri.001G053800.1	8.46	47692.5	1	Chr01:4100737..4104806	428
Potri.008G006500.1	5.34	46854.1	8	Chr08:377663..37976	422

Species name: *Selaginella moellendorffii*

gene name	pI	MW	chromosome location	genomic cordinates	peptide sizes

	pI	MW	chromosome location	genomic cordinates	peptide sizes
83932 putative glucuronosyltransferase (GLCAT)	9.5	47750.9	6	scaffold_6:2759488..2760956	422
74856 putative glucuronosyltransferase (GLCAT)	7.69	46361.1	0	scaffold_0:1673406..1674872	410

Species name: *Solanum lycopersicum*

gene name	pI	MW	chromosome location	genomic cordinates	peptide sizes
Solyc02g087550.2.1	9.19	49183.6	2	SL2.50ch02:49956733..49960809	430
Solyc03g077910.2.1	9.44	50225	3	SL2.50ch03:48527588..48532394	444
Solyc12g006490.1.1	8.93	47010.3	12	SL2.50ch12:1001341..1005831	412
Solyc07g054440.2.1	9.24	51232.3	7	SL2.50ch07:62748938..62756079	448
Solyc12g013850.1.1	8.26	48804.1	12	SL2.50ch12:4639057..4641430	426
Solyc09g010660.2.1	9.25	47634.4	9	SL2.50ch09:3982851..3987785	411
Solyc12g055960.1.1	6.84	44747.4	12	SL2.50ch12:61946358..61949889	388
Solyc10g007040.2.1	9.15	49287.9	10	SL2.50ch10:1431324..143409	435
Solyc10g079490.1.1	9.76	46226.6	10	SL2.50ch10:61027206..61032017	399
Solyc00g024670.1.1	9.41	44369.3	0	SL2.50ch00:12222064..12223921	384
Solyc05g008340.2.1	8.12	45543.1	5	SL2.50ch05:2699143..2703279	402
Solyc09g091230.2.1	6.57	48437	9	SL2.50ch09:70527340..70531224	427
Solyc05g053490.2.1	7.73	50056.5	5	SL2.50ch05:63569404..63575274	448
Solyc04g010290.2.1	8.21	45727.6	4	SL2.50ch04:3618701..3622866	401
Solyc02g081890.2.1	6.31	48714.3	2	SL2.50ch02:45663475..45665349	423

Species name: *Sorghum bicolor*

gene name	pI	MW	chromosome location	genomic cordinates	peptide sizes
Sobic.008G191500.1	6.62	47139.8	8	Chr08:62528298..62531964	425

gene name	pI	MW	chromosome location	genomic cordinates	peptide sizes
Sobic.010G184300.1	6.96	49070	10	Chr10:52411252..52415468	440
Sobic.001G131001.1	7.22	53535.4	1	Chr01:10290723..10301361	493
Sobic.003G027600.1	9.36	53402.6	3	Chr03:2429553..2431673	490
Sobic.007G036800.1	5.75	54437.8	7	Chr07:3329775..3333398	487
Sobic.003G027700.2	9.57	48706.8	3	Chr03:2437815..2441895	441
Sobic.001G418800.1	9.49	46551.4	1	Chr01:69959308..69963764	414
Sobic.001G506000.1	6.34	46047.5	1	Chr01:77401825..77410078	417
Sobic.003G092700.3	9.26	45180.6	3	Chr03:8060078..8064692	402
Sobic.009G046900.1	9.02	46767.8	9	Chr09:4444743..4447648	421
Sobic.003G311800.1	5.67	44488.1	3	Chr03:63993289..63995114	411

Species name: *Vitis vinifera*

gene name	pI	MW	chromosome location	genomic cordinates	peptide sizes
GSVIVT01020106001	9.48	45728.1	1	chr1:10281464..10283659	398
GSVIVT01015525001	9.11	53737.4	11	chr11:4492019..4499398	472
GSVIVT01014331001	9.33	47030.3	19	chr19:2716633..2720479	416
GSVIVT01030572001	7.11	37938.4	12	chr12:6838847..6843544	331
GSVIVT01034191001	9.13	32994.1	8	chr8:14537424..14541282	286
GSVIVT01032590001	6.49	27313.2	14	chr14:28625464..28631666	236
GSVIVT01011780001	8.97	45587	1	chr1:4209011..4214620	401
GSVIVT01029033001	8.5	47724.3	5	chr5:10988736..10994229	432
GSVIVT01029965001	9.78	31424.1	8	chr8:2432731..2436856	270

Species name: *Zea mays*

gene name	pI	MW	chromosome location	genomic cordinates	peptide sizes

gene name	pI	MW	chromosome location	genomic cordinates	peptide sizes
GRMZM2G423476_T01	7.79	51272.4	1	1:180715229..180718717	465
GRMZM2G150323_T01	7.3	49087.2	9	9:84966445..84971881	439
GRMZM2G043310_T01	9.11	60594.1	6	6:102760769..102765401	548
GRMZM2G063133_T01	8.9	49489.6	1	1:262568857..262571336	455
GRMZM2G055313_T01	9.72	50704.7	3	3:4706846..4710504	463
GRMZM2G038898_T01	10.38	48488.8	1	1:238522629..238525355	430
GRMZM2G178571_T01	6.92	46001.5	1	1:9906778..9917383	415
GRMZM2G060061_T03	9.05	45267.8	8	8:10809753..10813461	402
GRMZM2G111428_T01	5.51	43822.3	8	8:172044502..172046352	409
GRMZM2G018869_T01	8.23	45091.5	3	3:25216656..25220105	402

Species name: *Amborella trichopoda*

gene name	pI	MW	chromosome location	genomic cordinates	peptide sizes
AmTr_v1.0_scaffold00055.8	8.81	49016.1	NA	scaffold00055:214647..236380	434
AmTr_v1.0_scaffold00104.27	7.24	47927.1	NA	scaffold00104:840929..849452	426
AmTr_v1.0_scaffold00003.366	8.91	49359.8	NA	scaffold00003:8465763..8474471	435
AmTr_v1.0_scaffold00068.126	9.42	50136.1	NA	scaffold00068:2812995..2819646	443
AmTr_v1.0_scaffold00012.80	9.23	42137.6	NA	scaffold00012:2014751..2022762	369
AmTr_v1.0_scaffold00004.73	9.99	44205.5	NA	scaffold00004:1050285..1057590	384
AmTr_v1.0_scaffold00039.28	8.95	54669.8	NA	scaffold00039:691887..712985	488
AmTr_v1.0_scaffold00025.165	8.95	36867.2	NA	scaffold00025:2172974..2178885	323
AmTr_v1.0_scaffold00133.14	7.11	18790.4	NA	scaffold00133:446240..446737	165
AmTr_v1.0_scaffold00105.21	243	27970	NA	scaffold00105:413794..415020	243
AmTr_v1.0_scaffold00133.16	6.3	27970	NA	scaffold00133:449636..452065	267

Supplemental Table 2-2. Identification of Positively Selected Sites Using the Site Model, Branch Model and the Branch Site Model. For the branch model (A), bryophytes were used as foreground (branch I) while branches II and III corresponding to monocot species and dicot species respectively were used as background (See Materials and Methods; Figure 2D). The likelihood Ratio Test (LRT) p value were computed based on the comparison between the Model 0 (null hypothesis) vs. Two ratio Model 2 (alternative hypothesis). $P > 0.05$ suggest there are no sites preferentially under positive selection in branch I compared to other branches (branch II and III). Note that when monocots and dicots were made foreground, the LRT p value > 0.05 suggesting there are still no site under positive selection across branches II and III. For the Branch site model (B), Model A and Model A null were compared, the LRT P value < 0.05 indicating there are some sites that are potentially under positive selection. However the Bayes Empirical Bayes (BEB) computation were used to validate the potential positively selected sites identified in the branch site model and showed posterior probability estimates that were $< .95$ which is less than the cut off for a positively selected site ($PP>0.95$ to conclude that there is a positively selected site). For the site model analysis, only M7 vs. M8 model identified positively selected sites.

(A) Branch model (BM) using clade I
(bryophytes and lycophytes) as
foreground while II and III as backgrounds

Model	np	Ln L	Model compared	LRT P-value
Two ratio Model 2	28	-6739.459135		
Model 0	27	-6739.668291	Model 0 vs. Two ratio Model 2	0.517780527

Branch model (BM) for monocots, clade II as foreground while clade I and clade III as backgrounds

Model	np	Ln L	Model compared	LRT P-value
Two ratio Model 2	28	-6738.350642		
Model 0	27	-6739.668291	Model 0 vs. Two ratio Model 2	0.104512452

Branch model (BM) for dicot, clade III as foreground while clade I and II as backgrounds

Model	np	Ln L	Model compared	LRT P-value
Two ratio Model 2	28	-6739.537060		
Model 0	27	-6739.668291	Model 0 vs. Two ratio Model 2	0.608433751

(B) Branch site model (BSM) using clade I as foreground and clade II and III as background

Model	np	Ln L	Model compared	LRT P-value
Model A	30	-6812.739399		
Model A null	29	-6821.091640	Model A vs.Model A null	0.000043678

Branch site model (BSM) using clade II as foreground and clade I and III as background

Model	np	Ln L	Model compared	LRT P-value
Model A	30	-6812.128085		
Model A null	29	-6821.091640	Model A vs.Model A null	2.3E-05

Branch site model (BSM) using clade III as foreground and clade I and II as background

Model	np	Ln L	Model compared	LRT P-value
Model A	30	-6817.284229		
Model A null	29	-6821.091640	Model A vs.Model A null	0.005789

(C) Site model (SM); all 3 clades were jointly considered together

Model	np	Ln L	Model compared	LRT P-value
M3	31	-6665.388183		
M0	27	-6739.668291	M0 vs. M3	0
M2a	30	-6690.326318		
M1a	28	-6815.260596	M1a vs. M2a	0
M8	30	-6690.326318		
M7	28	-6821.091640	M7 vs.M8	0
M8a	29	-6815.270643	M8a vs.M8	0

Supplemental Table 3-1. List of Primers Used for Mutant Characterization

Purpose		
PCR	Sequence	
glcat14a-1_LP	TAGCCACACGACATGTTCAAG	
glcat14a-1_RP	ACTATGCTGCAGAATTTTTAAGTTC	
glcat14a-2_LP	ATTGGTTCAATCTTCGCTTTG	
glcat14a-2_.RP	TCAACCAATGAGAAATGGAGC	
glcat14c-1_LP	ACGACGTAATCGAATTGGATG	
glcat14c-1_RP	TGTAACGGTTGAGAATTTCCG	
LBb1.3	ATTTTGCCGATTTCGGAAC	
qPCR primers		
At5g39990.L1	ATTTGCCGCGGGATTTGAAC	
At5g39990.R1	CATCCAAGCAGAGCCTGTGA	
At5g39990.L2	GCTTTTCACAGGCTCTGCTTG	
At5g39990.R2	TGCAGGTCGCTATTTACCGT	
AT2G37585_L2	AGGCTTGAGAAGCTCATGGTT	
AT2G37585_R2	AGAACTCTTTAGACGAATGAGACCA	

Supplemental Table 4-1. List of Primers Used for Mutant Characterization

Purpose	
PCR	Sequence
glcat14a-1_LP	TAGCCACACGACATGTTCAAG
glcat14a-1_RP	ACTATGCTGCAGAATTTTTAAGTTC
glcat14a-2_LP	ATTGGTTCAATCTTCGCTTTG
glcat14a-2_.RP	TCAACCAATGAGAAATGGAGC
glcat14b-1_LP	TGATGTTTTCTGGGTTTCACAG
glcat14b-1_RP	AGTGACCATCAATAGGCCCTC
glcat14b-2_LP	GCAGCGTGAAGTGTATTAGCC
glcat14b-2_RP	TTGCGTGTTTAAGGGATTAGC
glcat14c-1_LP	ACGACGTAATCGAATTGGATG
glcat14c-1_RP	TGTAACGGTTGAGAATTTCCG
LBb1.3	ATTTTGCCGATTTCGGAAC
qPCR primers	
At5g39990.L2	GGATGATAACACCTGGCGGT
At5g39990.R2	AAGGCATGTTGTGTGGAGCA
At5g15050.L2	TCCGGGTGGTTGGTGTATTG
At5g15050.R2	GCCAAACTTTAGGTCGGCAA
At2g37585_L2	AGGCTTGAGAAGCTCATGGTT
At2G37585_R2	AGAACTCTTTAGACGAATGAGACCA

Supplemental Table 4-2. Amino Acid Sequence Similarity Matrix Among GT14 Family

	ATGLCAT14A	ATGLCAT14B	ATGLCAT14C	ATGLCAT14D	ATGLCAT14E	AT1G53100	AT1G03520	AT1G71070	AT3G03690	AT4G03340	AT4G27480
ATGLCAT14A		72.12	54.19	49.16	58.24	53.58	53.12	49.43	57.59	55.11	59.29
ATGLCAT14B			54.15	49	55.61	59.6	55.5	49.14	58.17	54.89	57.86
ATGLCAT14C				42.82	53.24	49.33	53.6	46.42	54.75	52.97	50.29
ATGLCAT14D					39.76	38.85	45.45	45	44.74	45.11	41.53
ATGLCAT14E						73.93	52.05	47.86	63.82	50.24	68.47
AT1G53100							56.45	44.7	58.77	50.6	64.82
AT1G03520								47.28	53.71	76.06	52.47
AT1G71070									47.94	43.77	42.52
AT3G03690										53.89	54.36
AT4G03340											52.26

Supplemental Table 4-3. Monosaccharide Composition Analysis of AGPs Extracted from Leaf Tissue of 40-day-old *glcat14* Mutants and WT. Values are relative to total sugar composition (expressed as mol %) of triplicate assays $\pm$ SE

	Fuc	Rha	Ara	Gal	Glu	Xyl	Man	GalA	GlcA
wild type	1.15 ± 0.39	9.35 ± 0.59	26.56 ± 0.57	36.60 ± 0.78	4.36 ± 0.57	2.67 ± 0.41	3.0 ± 0.8	11.20 ± 0.97	5.05 ± 0.42
glcat14a	1.05 ± 0.29	5.82 ± 0.16	27.84 ± 0.78	37.19 ± 0.21	5.18 ± 0.41	2.59 ± 0.15	3.87 ± 0.14	12.94 ± 0.89	3.48 ± 0.72
glcat14b	1.28 ± 0.21	9.16 ± 0.92	23.03 ± 0.53	40.59 ± 0.34	5.38 ± 0.57	3.99 ± 0.56	3.48 ± 0.15	10.28 ± 0.92	2.77 ± 0.35
glcat14c	0.93 ± 0.19	7.88 ± 0.63	23.44 ± 0.70	37.69 ± 0.68	8.65 ± 0.17	4.98 ± 0.26	4.17 ± 0.61	9.25 ± 0.16	2.97 ± 0.58
glcat14a/c	1.09 ± 0.72	5.51 ± 0.35	23.15 ± 0.12	46.35 ± 0.34	9.93 ± 0.48	2.04 ± 0.45	3.19 ± 0.38	5.85 ± 0.35	2.85 ± 0.27
glcat14b/c	2.45 ± 0.54	8.44 ± 0.69	24.85 ± 0.87	46.95 ± 0.71	3.77 ± 0.23	2.23 ± 0.14	3.13 ± 0.75	5.66 ± 0.81	2.47 ± 0.52
glcat14a/b	1.25 ± 0.39	3.87 ± 0.23	28.84 ± 0.45	46.69 ± 0.88	5.24 ± 0.32	1.85 ± 0.47	2.85 ± 0.88	8.23 ± 0.56	1.14 ± 0.12
glcat14a/b/c	1.67 ± 0.35	4.25 ± 0.29	26.34 ± 0.41	52.44 ± 0.47	4.34 ± 0.65	0.93 ± 0.2	1.96 ± 0.45	5.89 ± 0.44	2.15 ± 0.25

Supplemental Table 4-4. Monosaccharide Composition Analysis of AGPs Extracted from Stem Tissue of 40-day-old *glcat14* Mutants and WT. Values are relative to total sugar composition (expressed as mol %) of triplicate assays ± SE

	Fuc	Rha	Ara	Gal	Glu	Xyl	Man	GalA	GlcA
wild type	0.97 ± 0.27	2.16 ± 0.22	24.92 ± 0.59	53.33 ± 0.72	6.80 ± 0.34	1.28 ± 0.31	1.34 ± 0.4	5.05 ± 0.47	4.09 ± 0.49
glcat14a	0.87 ± 0.14	1.67 ± 0.17	24.65± 0.71	55.51 ± 0.41	4.26 ± 0.80	0.93 ± 0.24	1.06 ± 0.25	7.72 ± 0.48	3.28 ± 0.23
glcat14b	1.29 ± 0.24	1.51 ± 0.34	25.78 ± 0.34	57.48 ± 0.70	1.79 ± 0.58	0.86 ± 0.58	1.19 ± 0.26	6.89 ± 0.16	3.16 ± 0.26
glcat14c	0.92 ± 0.22	1.93 ± 0.14	26.47 ± 0.40	58.32 ± 0.71	2.22 ± 0.18	0.63 ± 0.25	1.15 ± 0.19	4.94 ± 0.28	3.37 ± 0.86
glcat14a/c	0.88 ± 0.43	1.72 ± 0.31	27.27± 0.36	53.30 ± 0.39	5.29 ± 0.14	3.21 ± 0.18	1.15 ± 0.17	3.83 ± 0.21	3.31 ± 0.28
glcat14b/c	0.91 ± 0.15	2.03 ± 0.53	28.99 ± 0.67	54.17 ± 0.75	3.76 ± 0.58	0.59 ± 0.17	1.30 ± 0.22	5.10 ± 0.26	3.12 ± 0.47
glcat14a/b	1.22 ± 0.28	1.05 ± 0.23	28.40 ± 0.48	56.72 ± 0.93	2.62 ± 0.13	1.11 ± 0.20	1.41 ± 0.23	4.77 ± 0.19	2.67 ± 0.22
glcat14a/b/c	1.33 ± 0.12	1.11 ± 0.25	25.74 ± 0.86	49.30 ± 0.88	11.66 ± 0.55	0.89 ± 0.17	0.68 ± 0.27	8.12 ± 0.37	1.11 ± 0.26

Supplemental Table 5-5. (A) Monosaccharide Composition Analysis of AGPs Extracted from Silique Tissue of 40-day-old *glcat14* Mutants and WT. Values are relative to total sugar composition (expressed as mol %) of triplicate assays ± SE; (B) Calcium analysis of AGPs extracted from silique tissue of 40-day-old *glcat14* mutants and WT (Col-0) expressed as percentage of calcium per AGPs, SE= Standard error

(A)

	Fuc	Rha	Ara	Gal	Glu	Xyl	Man	GalA	GlcA
wild type	1.15 ± 0.39	9.35 ± 0.59	26.56 ± 0.57	36.60 ± 0.78	4.36 ± 0.57	2.67 ± 0.41	3.0 ± 0.8	11.20 ± 0.97	5.05 ± 0.42
glcat14a	1.05 ± 0.29	5.82 ± 0.16	27.84 ± 0.78	37.19 ± 0.21	5.18 ± 0.41	2.59 ± 0.15	3.87 ± 0.14	12.94 ± 0.89	3.48 ± 0.72
glcat14b	1.28 ± 0.21	9.16 ± 0.92	23.03 ± 0.53	40.59 ± 0.34	5.38 ± 0.57	3.99 ± 0.56	3.48 ± 0.15	10.28 ± 0.92	2.77 ± 0.35
glcat14c	0.93 ± 0.19	7.88 ± 0.63	23.44 ± 0.70	37.69 ± 0.68	8.65 ± 0.17	4.98 ± 0.26	4.17 ± 0.61	9.25 ± 0.16	2.97 ± 0.58
glcat14a/c	1.09 ± 0.72	5.51 ± 0.35	23.15 ± 0.12	46.35 ± 0.34	9.93 ± 0.48	2.04 ± 0.45	3.19 ± 0.38	5.85 ± 0.35	2.85 ± 0.27
glcat14b/c	2.45 ± 0.54	8.44 ± 0.69	24.85 ± 0.87	46.95 ± 0.71	3.77 ± 0.23	2.23 ± 0.14	3.13 ± 0.75	5.66 ± 0.81	2.47 ± 0.52
glcat14a/b	1.25 ± 0.39	3.87 ± 0.23	28.84 ± 0.45	46.69 ± 0.88	5.24 ± 0.32	1.85 ± 0.47	2.85 ± 0.88	8.23 ± 0.56	1.14 ± 0.12
glcat14a/b/c	1.67 ± 0.35	4.25 ± 0.29	26.34 ± 0.41	52.44 ± 0.47	4.34 ± 0.65	0.93 ± 0.2	1.96 ± 0.45	5.89 ± 0.44	2.15 ± 0.25

(B)

genotype	(Ca/AGP) % ± SE
wild type	0.63 ± 0.14
glcat14a	0.64 ± 0.11
glcat14b	0.64 ± 0.19
glcat14c	0.65 ± 0.15
glcat14a/c	0.67 ± 0.29
glcat14b/c	0.67 ± 0.13
glcat14a/b	0.71 ± 0.31
glcat14a/b/c	0.77 ± 0.29

Appendix B: Supplemental Figures

Note: The supplemental figures are numbered according to the chapter they appear.

Supplemental Figure 2-1. Protein Sequence Alignment of Characterized GT14 Sequences from Arabidopsis, Humans, Mouse, Rat, *C. elegans, Gallus gallus, D. melanogaster*. The N terminal region (Black boxed region) and the C terminal (green boxed region) DXD motif has been functionally characterized in human XT-1; only the C terminal (green boxed region) DXD motif is utilized for its catalytic activity as resolved by site directed mutagenesis study [24]. This functionally important motif is absent in plants. Red boxed region indicated an uncharacterized DXD motif with a conserved tryptophan residue shared by all species. Asterisk indicate conserved W.

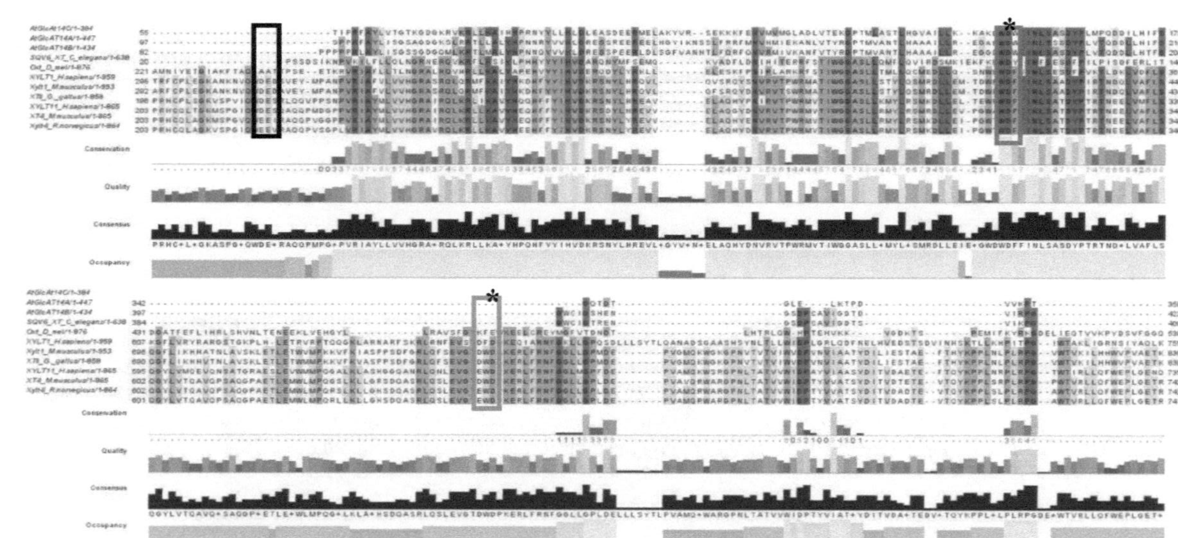

Supplemental Figure 2-2. Map Showing the Chromosomal Locations of *GLCAT* Genes in *Arabidopsis thaliana*, *Arabidopsis lyrata*, *Glycine max*, *Gossypium raimondii*, *Populus trichocarpa*, *Solanum lycopersicum*, *Sorghum bicolor*, *Brachipodium distachyon*, *Vitis vinifera* and *Oryza sativa*. The number of genes on each chromosome are indicated.

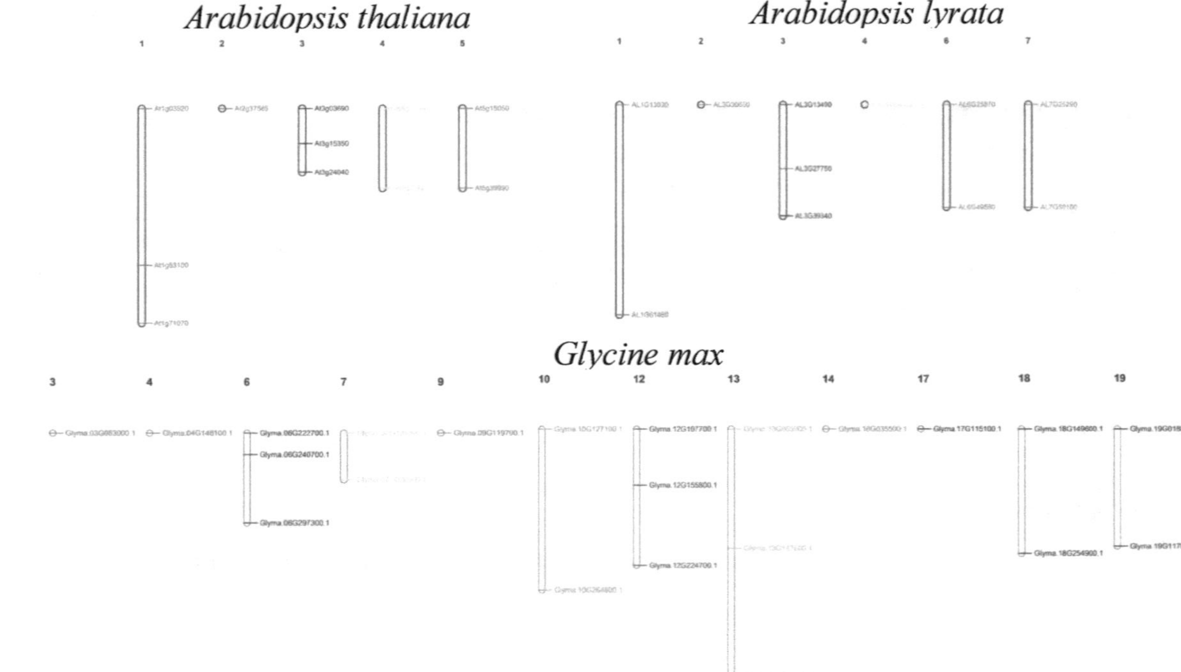

Supplemental Figure 2-2: continued

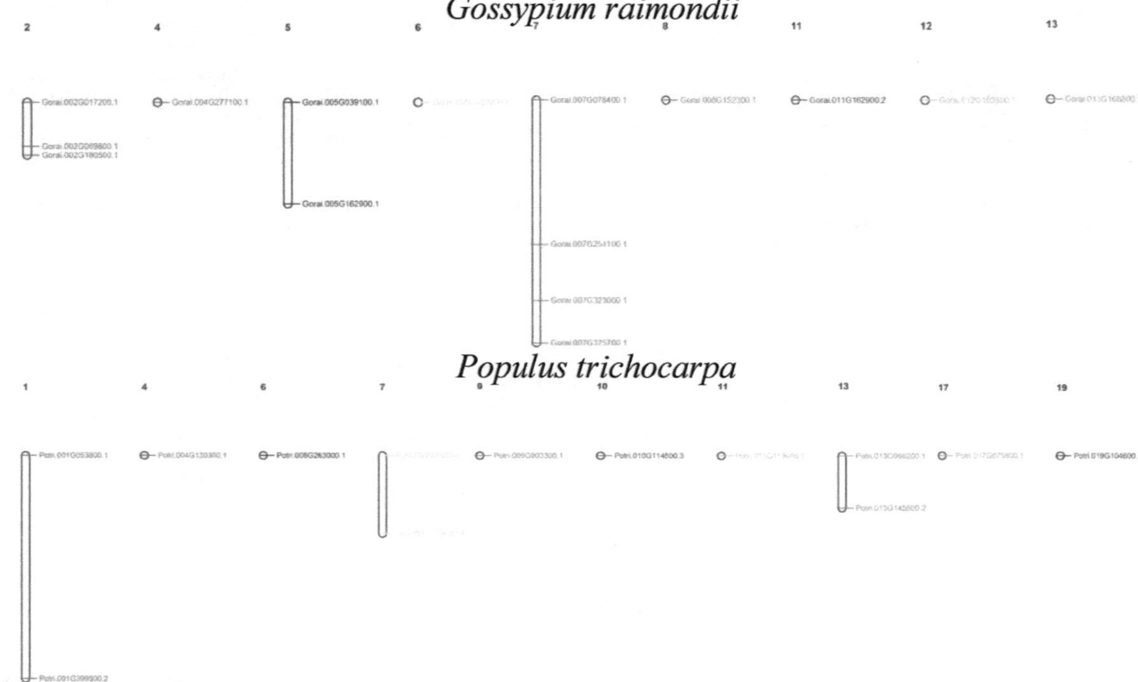

Supplemental Figure 2-2: continued

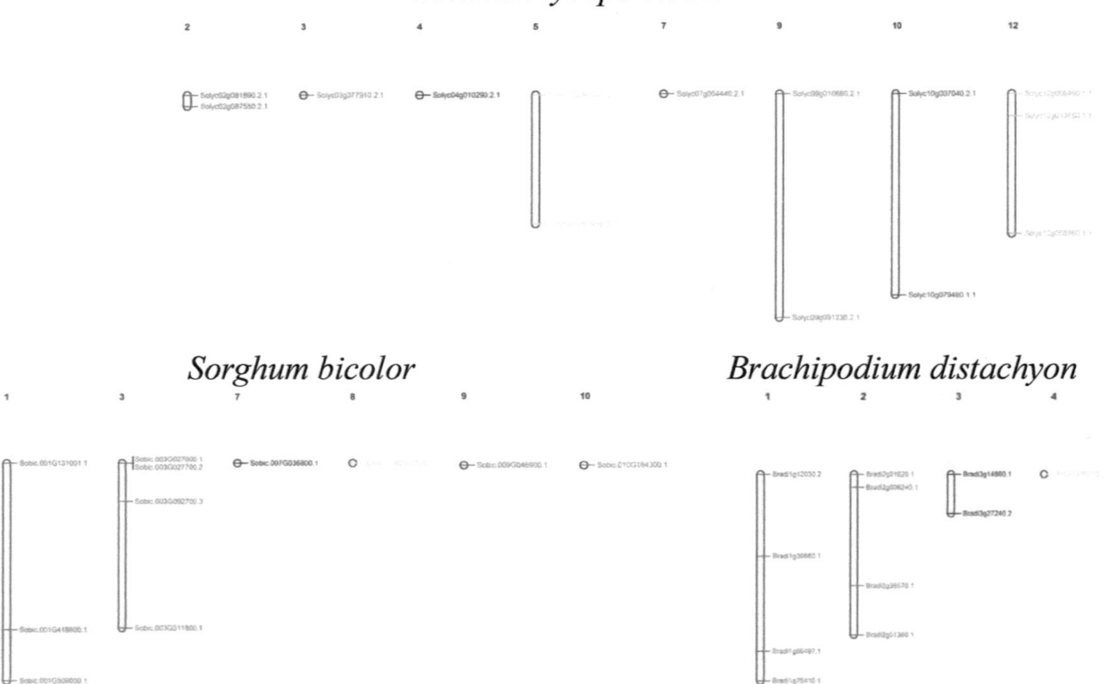

Supplemental Figure 2-2: continued

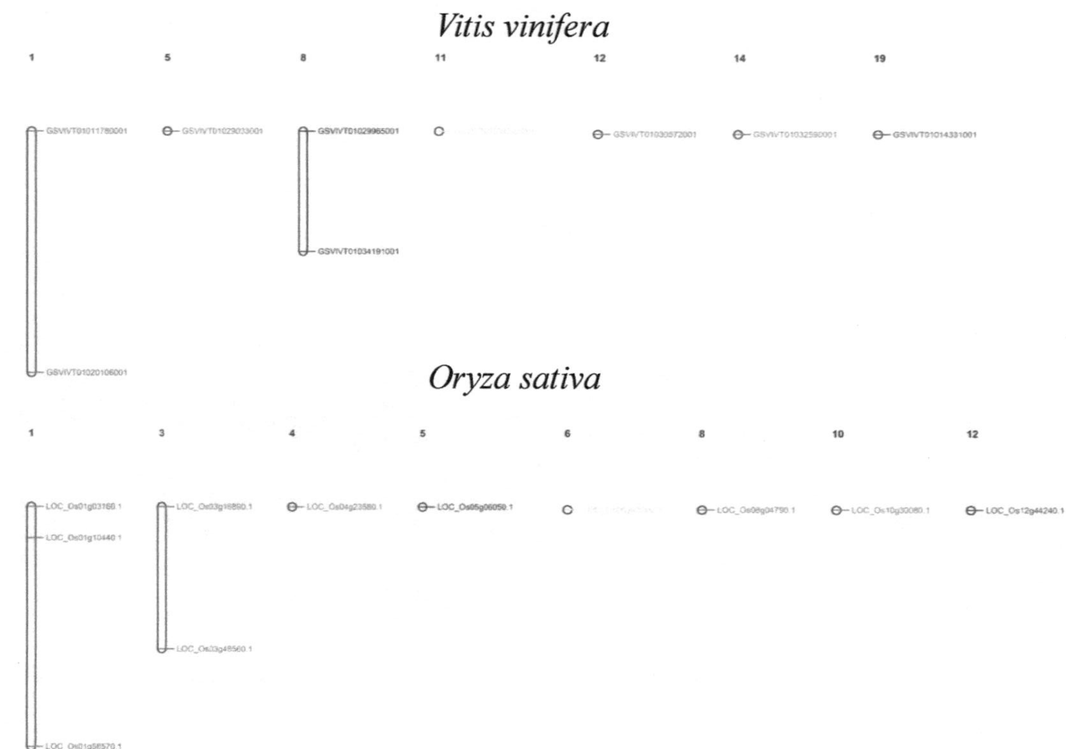

Supplemental Figure 2-2: continued

Zea mays

Supplemental Figure 2-3. Gene Structure Display of *GLCAT* Sequences from Various Plant Species. Gene structure is representative of bryophytes, lycophytes and dicots (A) and monocots (B) as illustrated by using the GSDS 2.0 server.

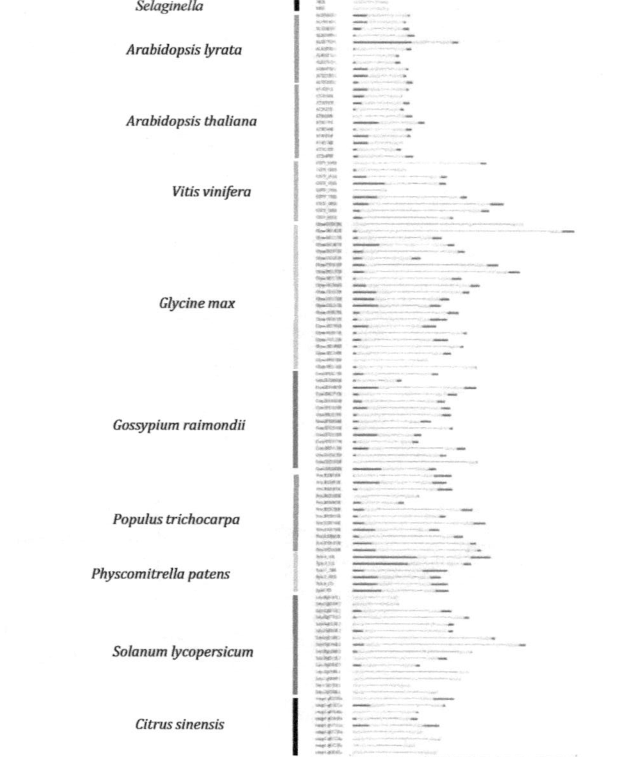

Supplemental Figure 2-3: continued

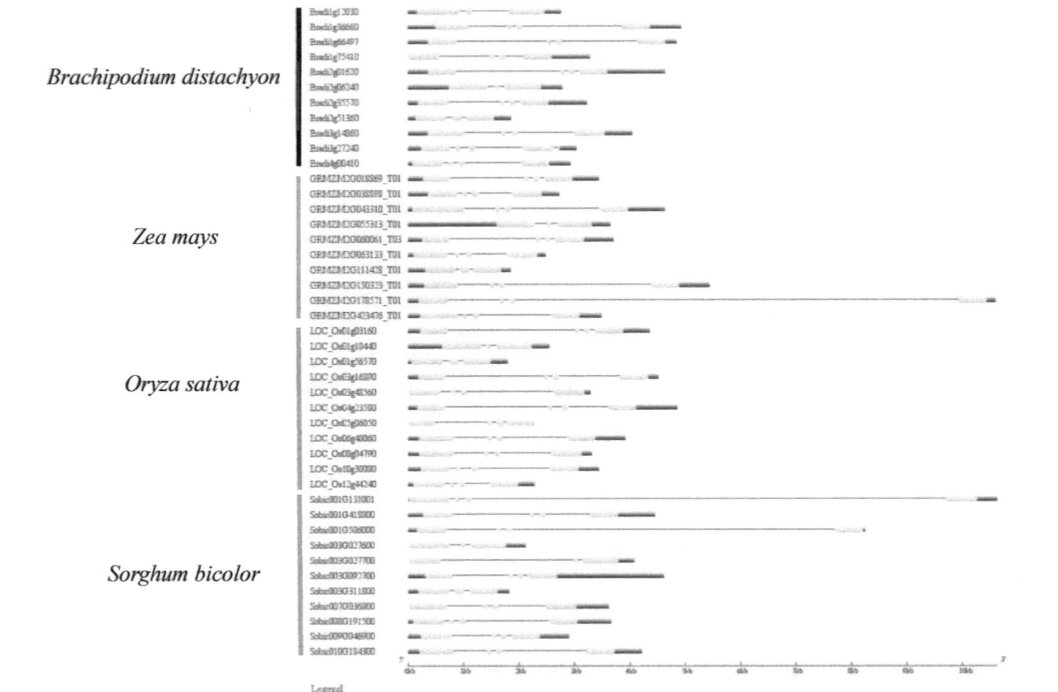

Supplemental Figure 2-4. Synteny Analysis of *GLCAT* Gene Families in Plant Species. Synteny analysis of *Arabidopsis thaliana* (A) *A lyrata*, (B) Brachipodium, (C) Citrus, (D) Soybean, (E) Cotton, (F) Rice, (G) *P patens*, (H) Poplar, (I) Selaginella, (J) Tomato, (K) Sorghum and (L) *Vitis vinifera*.. Inside the circle, ribbons represent local alignments based on bit score, red (> 80%), orange (> 60%), green (> 40%) and blue (>20%). Ribbon width is correlated with % identity. Ribbons representing best hits are outlined and placed on top of all other ribbons. Histogram on the top of the ideograms, shows how many times each colour has hit the specific part of the sequence.

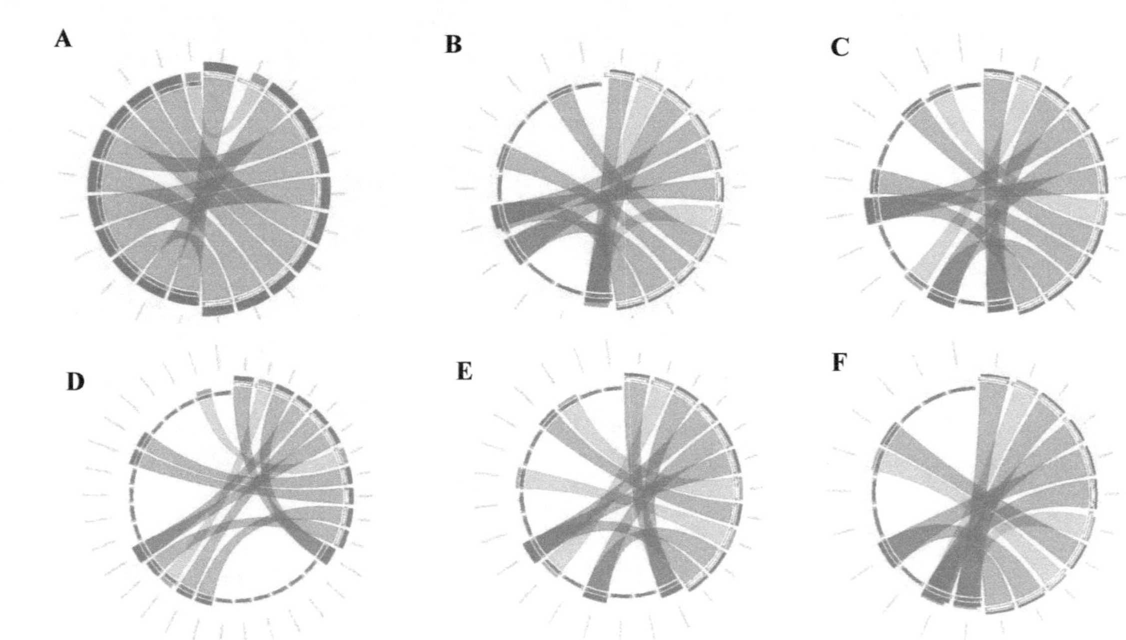

Supplemental Figure 2-4: continued

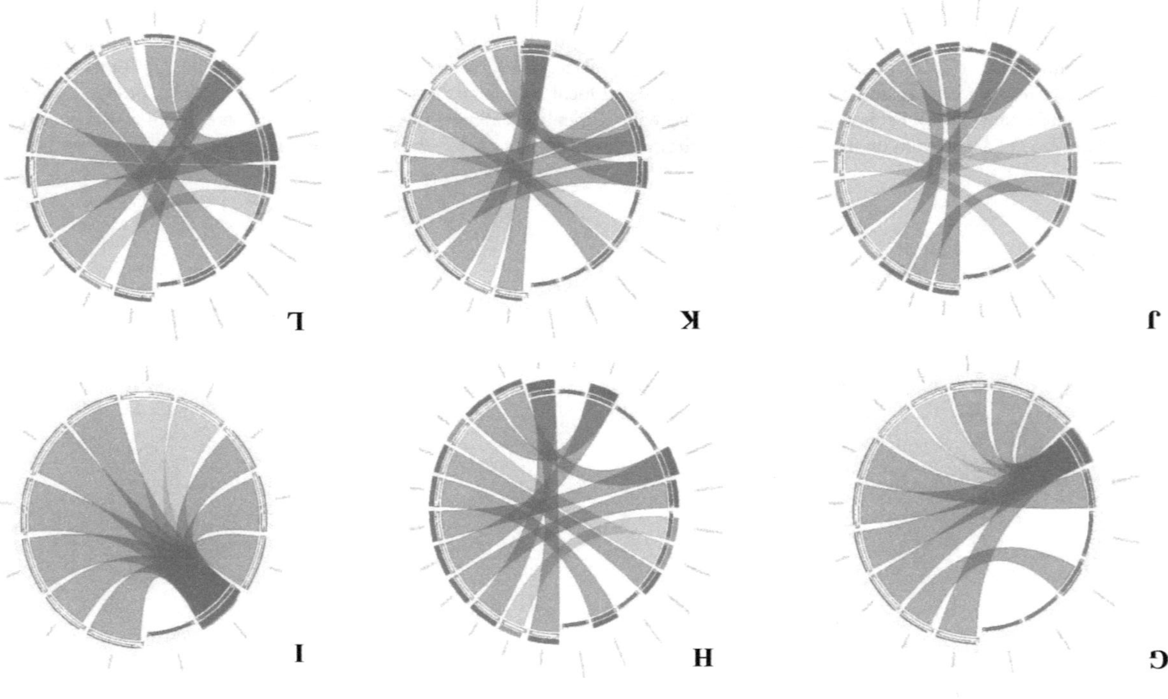

Supplemental Figure 2-5. Conserved Motifs/Blocks as Predicted by MEME Software. These eight identified motifs are found in 95% of the GLCAT sequences derived from multiple sequence alignment of all GLCAT family members from the 14 plant genomes examined.

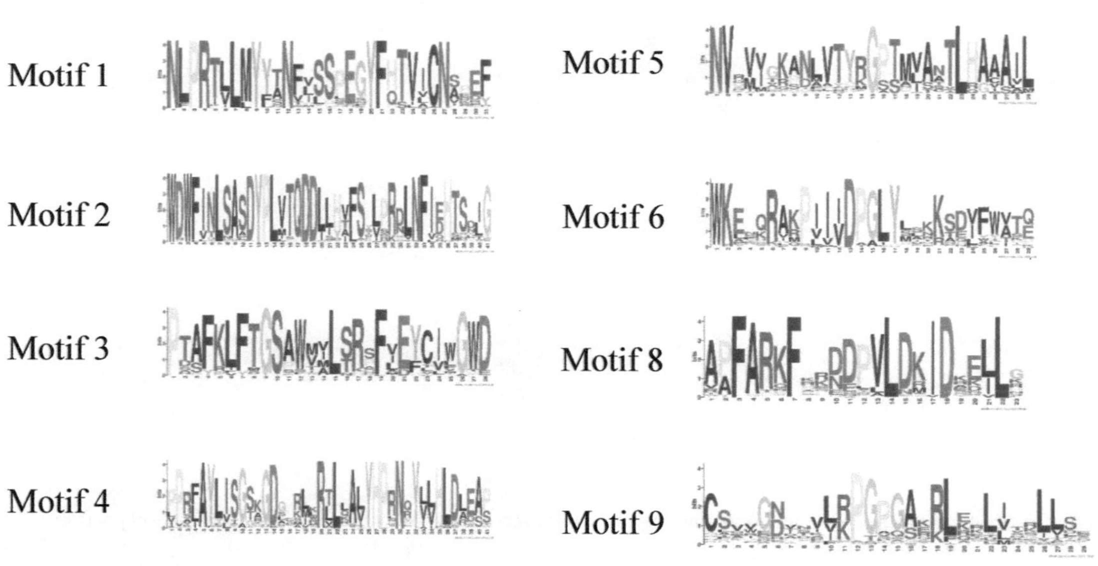

Motif 1

Motif 2

Motif 3

Motif 4

Motif 5

Motif 6

Motif 8

Motif 9

Supplemental Figure 2-6. Phylogenetic Analyses of GLCATs in Plant Genomes Using Maximum Likelihood Method. Figure showed the phylogenetic analyses in 14 plant species investigated in addition to putative GLCAT sequences in Amborella genome (indicated with a purple color). The eleven putative GLCAT sequences identified in the *Amborella trichopoda* genome occupy distinct nodes and notably evolved prior to the common ancestor of angiosperms. Two main evolutionary events were discovered and depicted with red and blue clades.

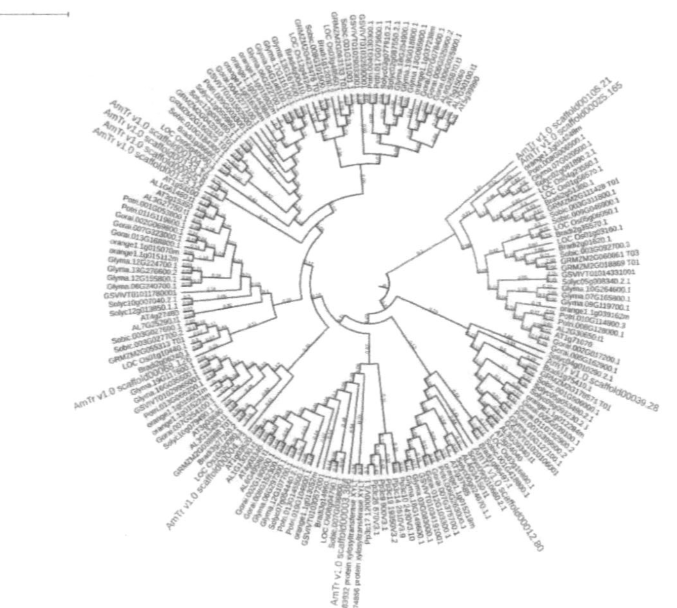

Supplemental Figure 3-1. Expression of the *GLCAT14A*, *GLCAT14B*, *GLCAT14C*, *FEI2* and *SOS5* Genes in the Arabidopsis Seed Coat. The Arabidopsis eFP browser (http://bar.utoronto.ca/efp_seedcoat/cgi-bin/efpWeb.cgi) was used to examine the expression of the *GLCAT* genes involved in glucuronidation of Type II AG and was compared to the expression of the *FEI2* and *SOS5/FLA4* genes in the seed coat at 3 day post anthesis (DPA), 7 DPA and 11 DPA.

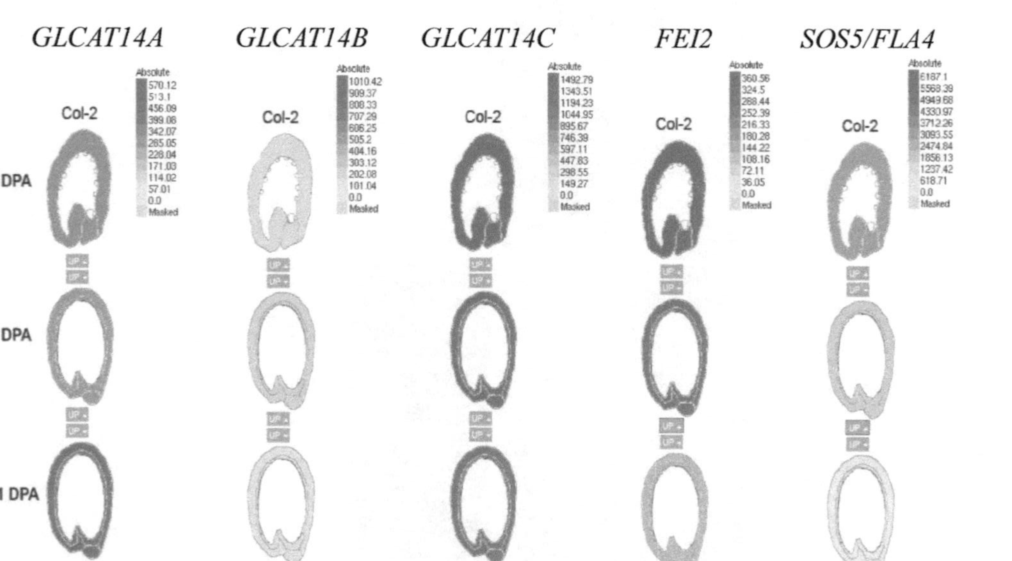

Supplemental Figure 3-2. Mucilage Phenotypes of WT and *glcat14* Mutants Hydrated in Different Chemical Extractants. A-D, Staining of the adherent mucilage with 0.01% ruthenium red (RR) after vortexing briefly for 5 min in water (A-D), Na_2CO_3 (E-H) and 50mM $CaCl_2$ (I-L). WT (M) and *glcat14a-1glcat14c-1* (N) seeds were shaken in water and stained with calcofluor, which primarily stains cellulose. Images (A-L) were acquired using light microscope, while images M and N were acquired using a Zeiss confocal microscope using the same acquisition settings to acquire both images. Three independent experiments (each with more than 25 seeds) were performed with similar results. Bar = 200µm for (A-L); Bar = 100 µm (M and N)

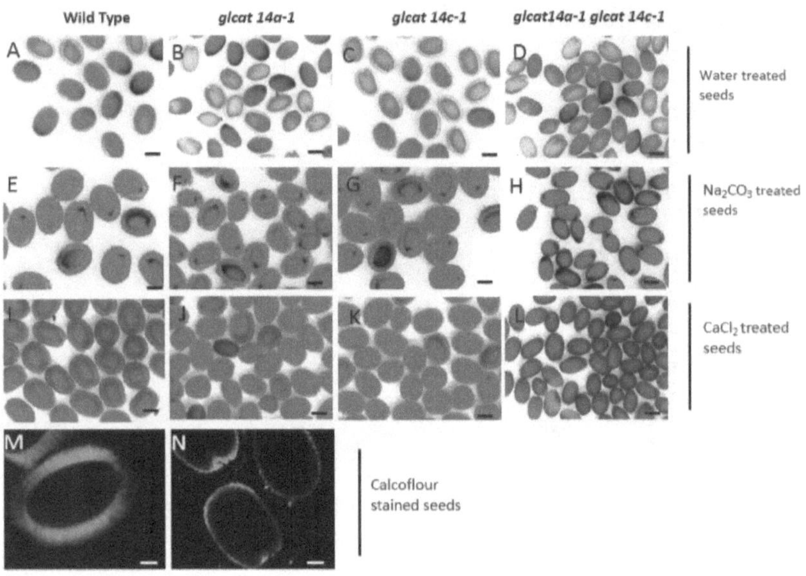

Supplemental Figure 3-3. Immunolabeling of Crystalline and Amorphous Cellulose in WT and *glcat14* Mucilage. Immunolabeling of crystalline cellulose with the CBM3a antibody (A) which binds preferentially to crystalline cellulose and the CBM28 antibody (B) which binds preferentially to amorphous cellulose in the adherent mucilage of the WT and *glcat14* mutants. The cellulosic ray-structure was counterstained with the S4B dye (red fluorescence). Three independent experiments (each with more than 25 seeds) were performed and similar results were obtained in each case. All scale bars = 50 μm.

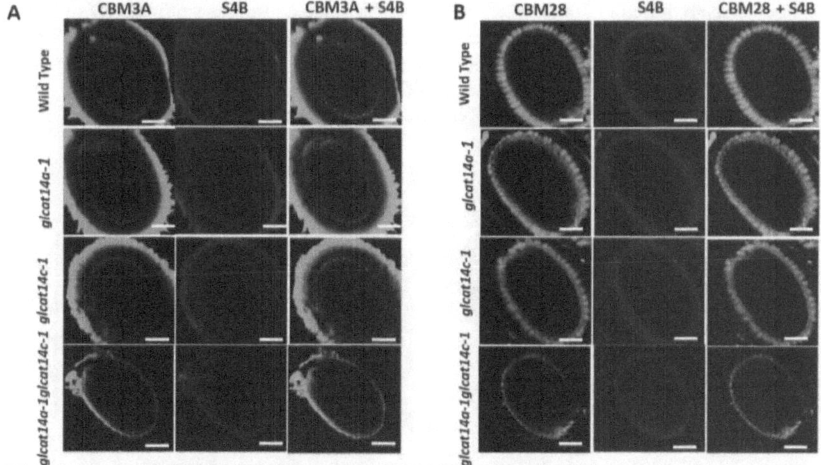

Supplemental Figure 3-4. Mucilage Phenotypes of WT and *glcat14* Mutants. A-D, Staining of the adherent mucilage with 0.01% ruthenium red (RR) after vortexing briefly for 5 mins in 50mM EDTA (A-H) and 50mM $CaCl_2$ (I-P). E-H represents higher magnification of A-D, while M-P represents higher magnification of I-L. Lack of detectable adherent mucilage were observed for both EDTA (D and H) and $CaCl_2$ imbibed (L and P) double mutant seeds. Addition of 50mM $CaCl_2$ was unable to rescue the mucilage defect in *glcat14a-1 glcat14c-1* mutants seeds (L and P). Images (A-P) were acquired using light microscope with the same acquisition settings. Quantification of the average area of ruthenium red stained mucilage capsule for WT and *glcat14* mutant seeds hydrated in 50mM $CaCl_2$(Q). The mucilage capsule is significantly reduced in *glcat14a-1* while *glcat14a-1glcat14c-1* could not be detected. Box plots were generated from 3 biological replicates of (>20 seeds each). The single and double asterisk marks a significant decrease compared with WT (Student's *t*-test, $P < 0.05$ for single asterisks and $P < 0.01$ for double asterisks). ND- Not determined. Bar = 100µm

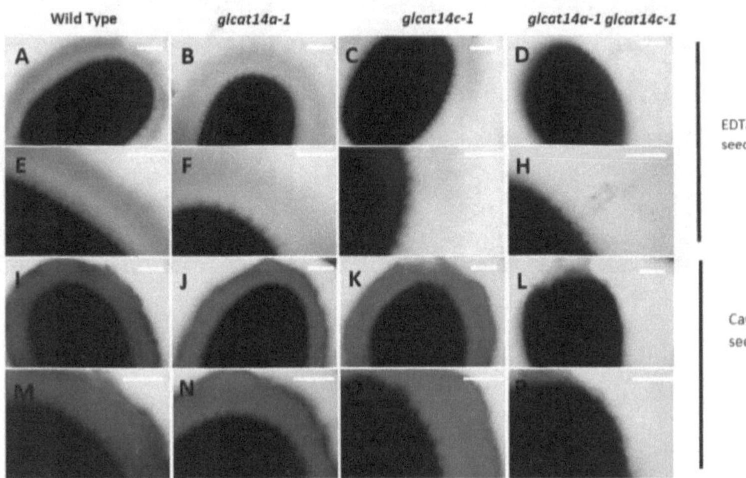

Supplemental Figure 3-4: continued

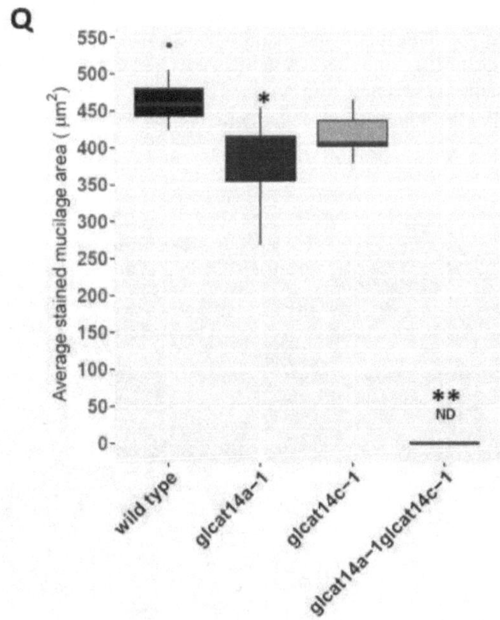

Supplemental Figure 3-5. Pectin Immunolabeling in the WT and *glcat14* Mucilage. Immunolabeling of WT and *glcat14* adherent mucilage with the CCRC-M35 antibody (A1-L1), JIM5 (A2-L2), JIM7 (A3-L3) and JIM13 (A4-L4) counterstained with S4B (red fluorescence). CCRC-M35 binds to unsubstituted rhamnogalacturonan I while JIM5 and JIM7 bind to partially methylesterified and highly methylesterified pectins, while JIM13 recognizes carbohydrate moieties associated with AGPs. Three independent experiments (each with more than 25 seeds) were performed and similar results were obtained. Scale bars = 50 μm, for A1-L1; Scale bars = 100 μm

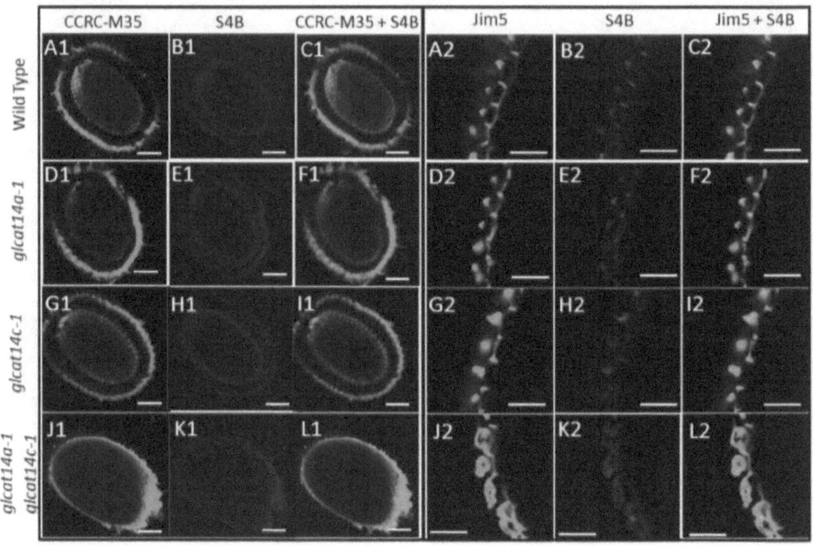

Supplemental Figure 3-5: continued

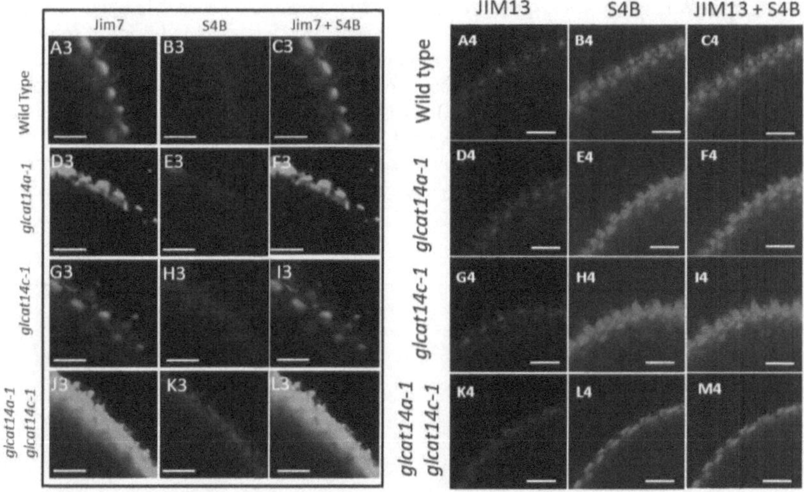

Supplemental Figure 3-6. Pectin Immunoblotting of Extracted Mucilage of WT and *glcat14* Mutant Seeds. Water-soluble and adherent mucilage was sequentially extracted from WT, *glcat14a-1*, *glcat14c-1*, *glcat14a-1glcat14c-1* seeds. Mucilage was diluted in a series of concentrations (for CCRC-M35 and JIM7) as specified prior to spotting on to nitrocellulose membrane. The membrane was hybridized with antibodies specifically binding to the unbranched RG I backbone (CCRC-M35, A), and antibodies specific to pectin HG (JIM5, B and JIM7, C). CCRC-M35 enzyme-linked immunosorbent assay (ELISA) of non-adherent and adherent mucilage showed increased CCRC-M35 epitopes for *glcat14a- 1* and *glcat14c-1* mutants, and a reduced epitope binding for *glcat14a-glcat14c-1* double mutants. The y-axis denotes the CCRC-M35 ELISA-corrected absorbance

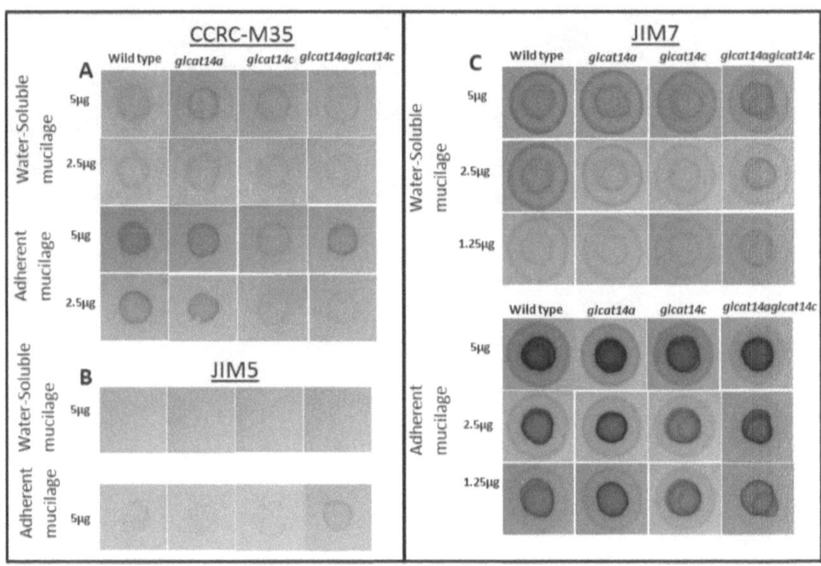

Supplemental Figure 3-6: continued

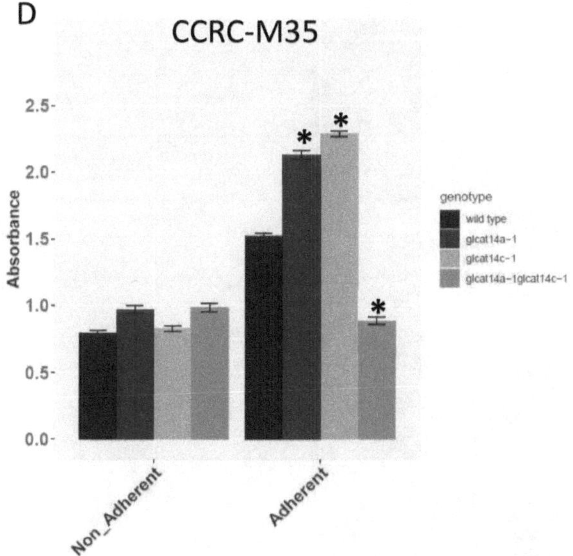

Supplemental Figure 3-7. Scanning Electron Microscopy of WT, *glcat14a-1*, and *glcat14c-1* Seeds. SEM of the seed coat surface of *glcat14a-1* and *glcat14c-1* mutant seeds are comparable to the WT.

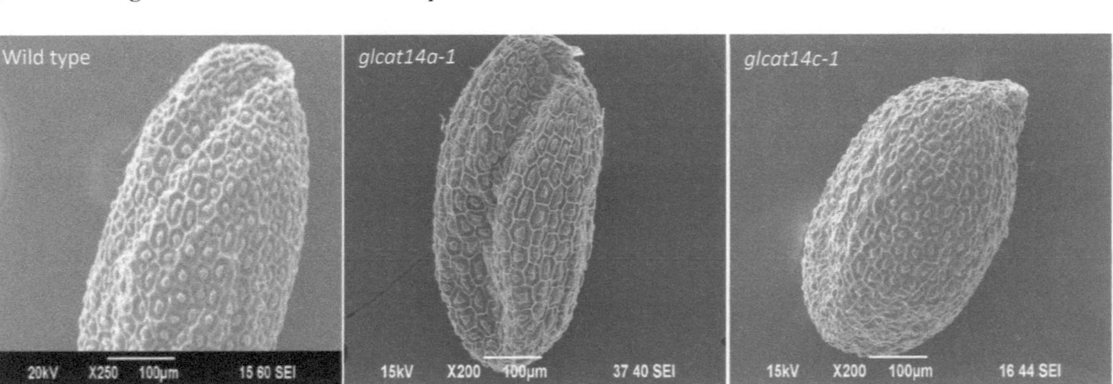

Supplemental Figure 4-1. PCR Screening of *glcat14* T-DNA Insertion Mutants. PCR screening of single mutants (A) and *glcat14a-1glcat14c-1* double mutant (B) and *glcat14b-1glcat14c-1* double mutant (C). A1, A2, B1, B2, C1 corresponds to *glcat14a-1*, *glcat14a-2*, *glcat14b-1*, *glcat14b-2*, *glcat14c-1* single mutants respectively; LR – Left and Right primers were used for PCR screening, BR – border primer and right primer were used for PCR screening. Red rectangles in panels B and C represents the confirmed homozygous double mutants, lane 2 in panel B was confirmed homozygous *glcat14a-1 glcat14c-1* mutants, while lane 4 in panel C was confirmed homozygous *glcat14b-1 glcat14c-1* mutants; M- Molecular ladder.

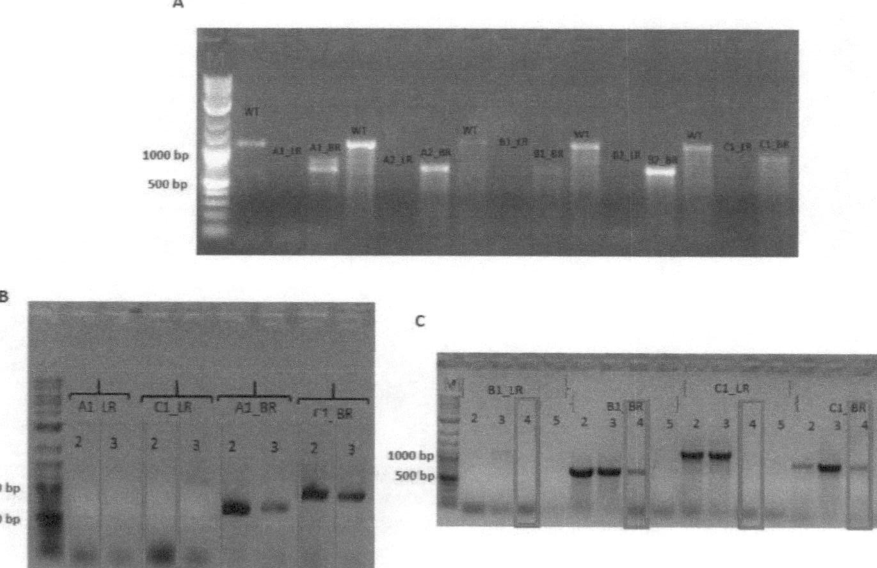

Supplemental Figure 4-2. Growth Measurements of *glcat14* Mutants and WT. A. Root length phenotypes of WT and *glcat14* mutants. Results showed that the *glcat14* mutants were comparable to WT. B Length of roots and hypocotyls from seedlings grown in the dark for 5 d was compared to WT as 100% (n > 40). No significant differences were observed between WT and *glcat14* mutants.

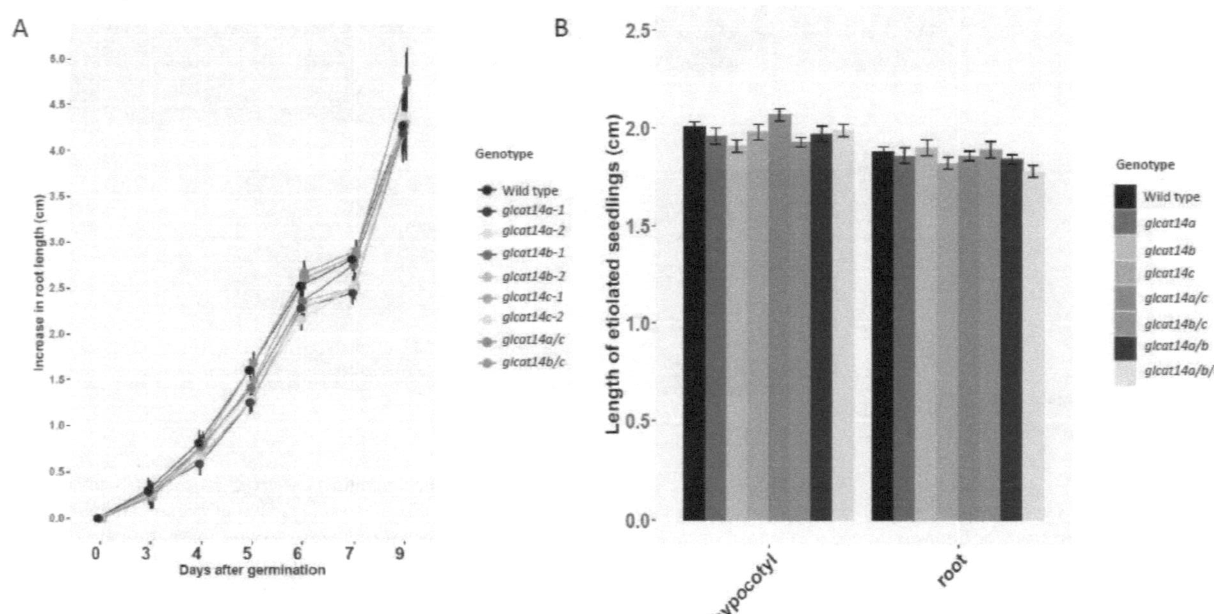

Supplemental Figure 5-1. SEM Images of Pollen Grains of WT, *glcat14a/b* and *glcat14a/b/c* Mutants. Surface view of SEM images of pollen grains of WT (A), *glcat14a/b* (B) and *glcat14a/b/c* (C). Red arrow in *glcat14a/b* and *glcat14a/b/c* indicated exine fusion while inset in C, displayed an enlarged image of pollen exine fusion. SEM images of pollen grains of WT (D), *glcat14a* (E), *glcat14b* (F), *glcat14c* (G), *glcat14a/c* (H)), *glcat14b/c* showed similarity with the WT.

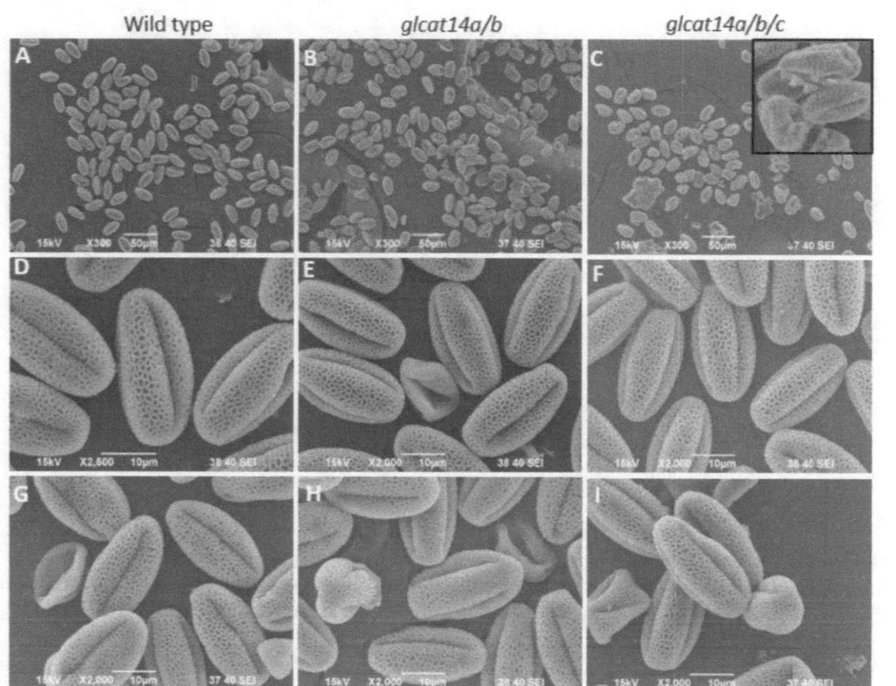

Supplemental Figure 5-2. TEM Micrographs Reveal Defects in Pollen Development in *glcat14a/b* and *glcat14a/b/c* Mutants. TEM micrographs of longitudinal sections of resin-embedded anthers of mature pollen grains of WT (A), *glcat14a/b* (B, C) and *glcat14a/b/c* (D-F) mutants. Extensive cytoplasmic shrinkage and eventual collapse of pollen grains were observed in *glcat14a/b* and *glcat14a/b/c* mutants. Scale bars = 2μm in (A, B, D, E), 10μm in (C, F).

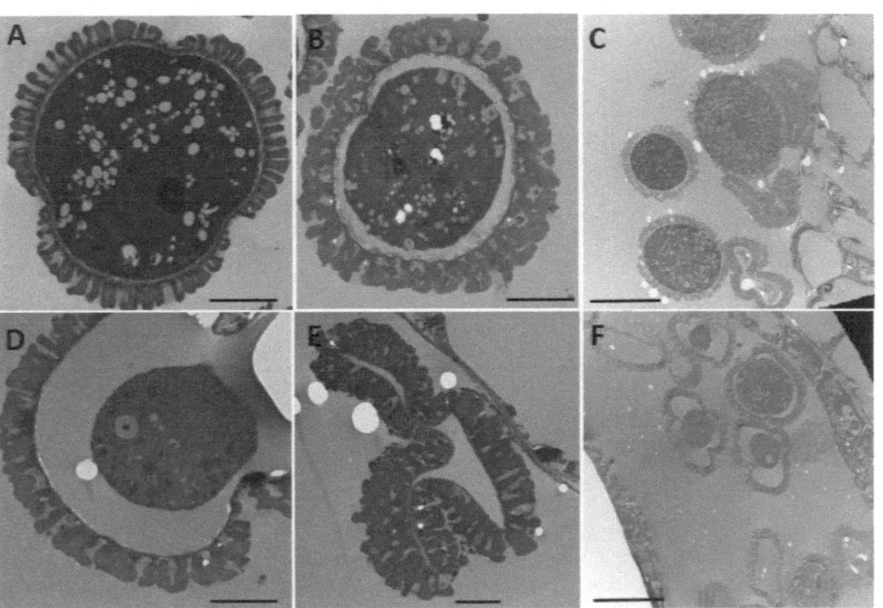

Supplemental Figure 5-3. Distribution of JIM13-epitope Labeling of AGPs of Arabidopsis Anthers in WT, *glcat14a/b* and *glcat14a/b/c* Mutants. Cross sections of resin-embedded anthers of WT and glcat14 mutants at bicellular stage were labelled with JIM13 and subsequently with Alexa Fluor 488-labeled secondary antibody and results were comparable to the WT except for the increased pollen collapse in *glcat14a/b* and *glcat14a/b/c* mutants. White arrows in a, showed JIM13 immunolabeling of the male gametes. En, Endothecium; M, microspore; Tp, tapetal cells. Bars = 100μm.

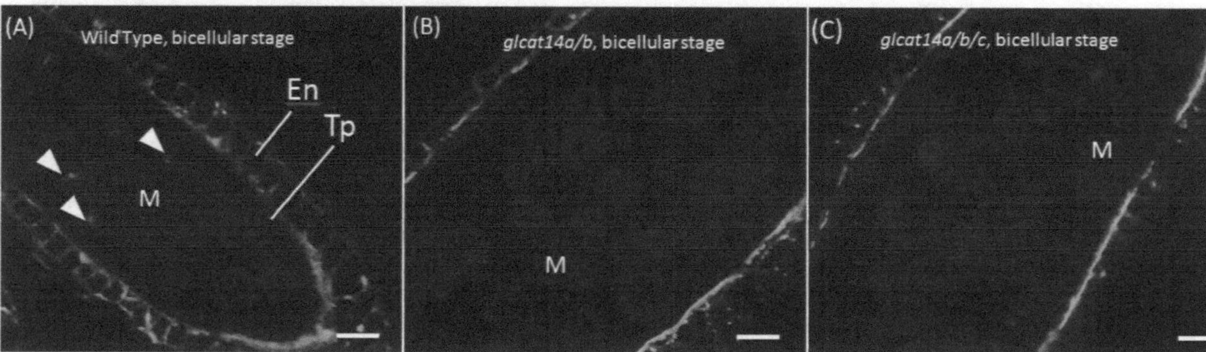

Supplemental Figure 5-4. Distribution of JIM5 and JIM7-epitope Labeling of Arabidopsis Anthers at Different Stages of Pollen Development in WT and *glcat14a/b and glcat14a/b/c* Mutants. Cross sections of resin-embedded anthers at the uninucleate stage, were labelled with JIM5 and showed weak labelling of developing microspores in WT (A) and *glcat14a/b/c* mutants (B); no differences were observed in JIM5 immunolabelling of WT and *glcat14* mutants. JIM7 immunolabelling of resin-embedded anther cross sections at the uninucleate (WT, C; *glcat14a/b* mutant, E) and bicellular stages (*glcat14a/b* /c mutant, F). Loss of JIM7 labelling in developing microspores of non-viable pollen grains (white asterisks in F) in *glcat14a/b/c* mutants were observed, while inset in D, showed pollenless locules in *glcat14a/b/c* mutants as seen in Figure 3. Scale bar = 200 μm, except for F which is 500 μm.

JIM5　　　　　　JIM7　　　　　　JIM7

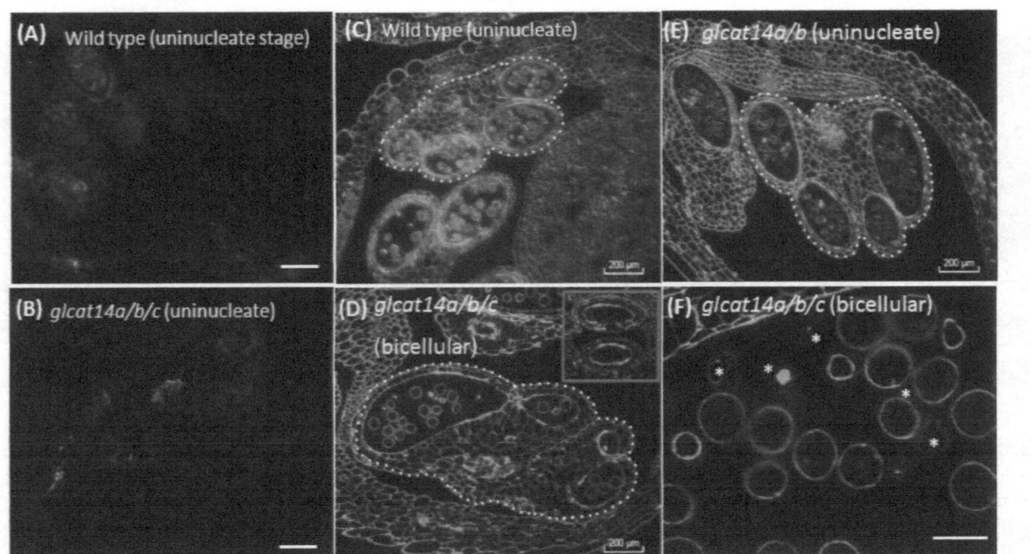

Supplemental Figure 5-5. Silique Length and JIM5 Immunolabeling of Pollen Tube of WT and *glcat14* Mutants. (A) Images of siliques of WT and *glcat14a/b/c* mutants and statistical comparisons (box plots) (B). S1-S15 in b, represents the first 15 siliques. Asterisks show significant differences between WT and *glcat14a/b/c* mutants (student's *t*-test, $P < 0.01$, $n \geq 25$ siliques for S1-S15 per genotype (C) JIM5 immunolabelling of WT and *glcat14* mutants. Immunolabelling of *glcat14* mutants were comparable to the WT. (D) *In-silico* gene expression analyses of *GLCAT14A*, *GLCAT14B* and *GLCAT14C* genes during seed developmental stages (Winter et al., 2007). Black arrow points to the micropyler endosperm near the filiform apparatus. Scale bar in A= 1cm; C = 100 μm.

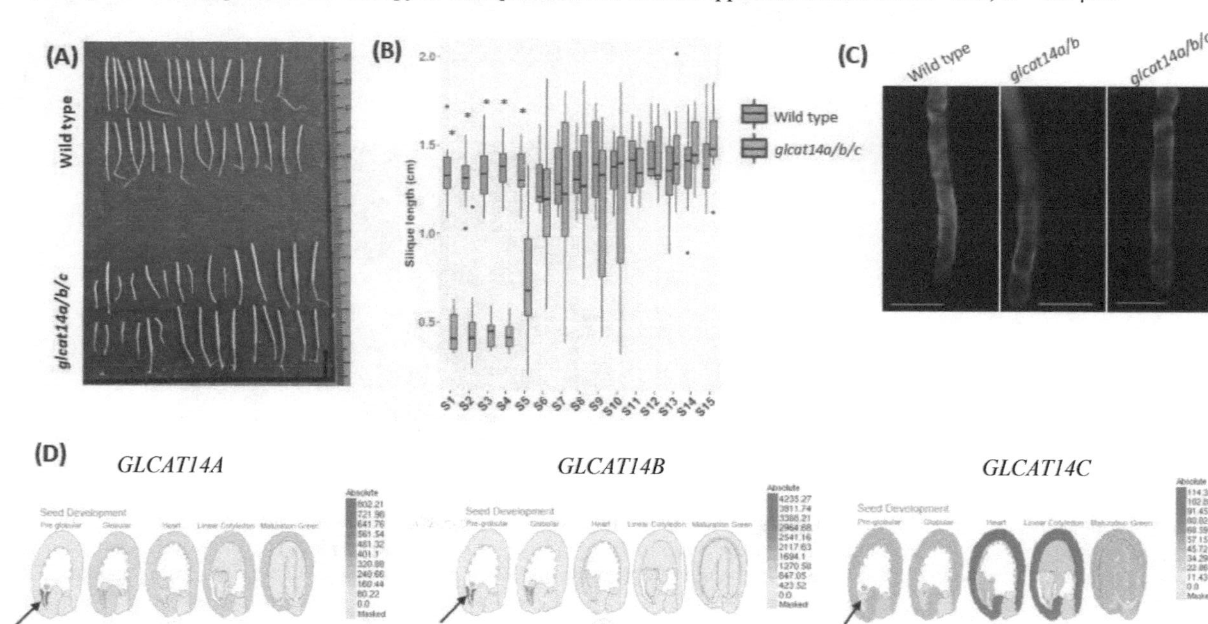

Supplemental Figure 5-6. Vanillin Staining of WT and *glcat14* Mutant. Vanillin-stained WT and *glcat14* ovule at 3 DAP under low, A and high magnification, B. Scoring of vanillin staining in WT, *glcat14a/b* and *glcat14a/b/c* mutant ovules were categorized in C and the percentage of stained ovules for each genotype were plotted in D. Numbers on top in D, indicate total ovules counted. Scale bar in A panel = 5mm; B, C panel =100 µm

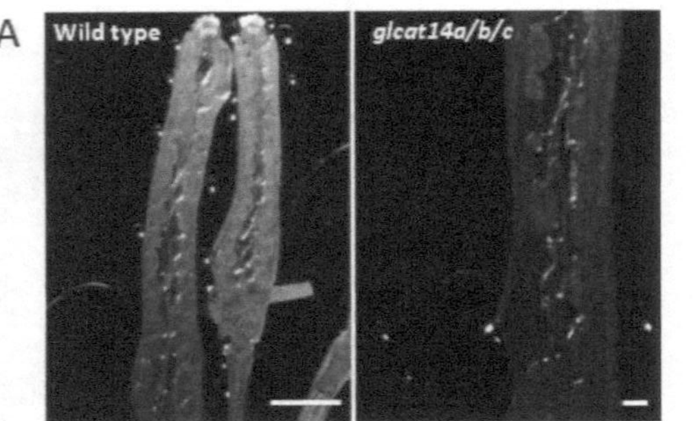

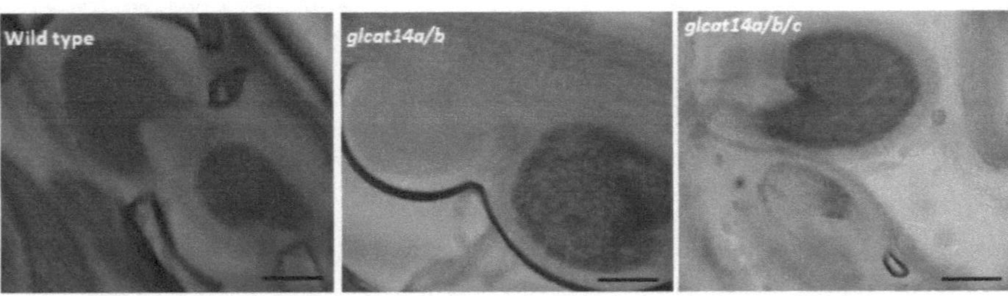

Supplemental Figure 5-6: continued

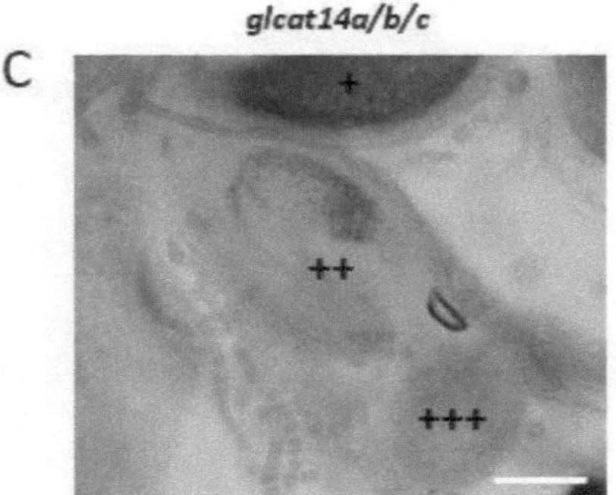

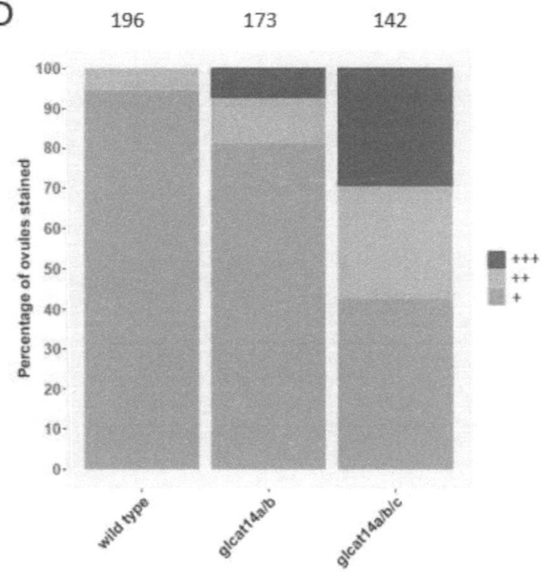

www.ingramcontent.com/pod-product-compliance
Lightning Source LLC
LaVergne TN
LVHW041453170726
843492LV00005B/1217